INTRODUCTION TO COMPUTER DESIGN AND IMPLEMENTATION

DIGITAL SYSTEM DESIGN SERIES

ARTHUR D. FRIEDMAN, Editor
George Washington University

S. I. AHMAD AND K. T. FUNG
Introduction to Computer Design and Implementation

MICHAEL ANDREWS
Principles of Firmware Engineering in Microprogram Control

J. E. ARSENAULT AND J. A. ROBERTS
Reliability and Maintainability of Electronic Systems

JEAN-LOUP BAER
Computer Systems Architecture

MELVIN A. BREUER
Digital System Design Automation:
Languages, Simulation and Data Base

MELVIN A. BREUER AND ARTHUR D. FRIEDMAN
Diagnosis and Reliable Design of Digital Systems

ARTHUR D. FRIEDMAN
Logical Design of Digital Systems

ARTHUR D. FRIEDMAN AND PREMACHANDRAN R. MENON
Theory and Design of Switching Circuits

INTRODUCTION TO COMPUTER DESIGN AND IMPLEMENTATION

S. IMTIAZ AHMAD and KWOK T. FUNG
University of Windsor

COMPUTER SCIENCE PRESS

Printed in the United States of America.

Computer Science Press, Inc.
11 Taft Ct.
Rockville, Maryland 20850

Printing 2 3 4 5 85 84 83 Year

Library of Congress Cataloging in Publication Data

Ahmad, Syed Imtiaz.
Introduction to computer design and implementation.

(Digital system design series)
Bibliography: p.
Includes index.
1. Electronic digital computers—Design and construction.
I. Fung, Kwok T., joint author. II. Title. III. Series.
TK7888.3.A37 621.3819′582 80-21005
ISBN 0-914894-11-0

Cover design by Ruth Ramminger

PREFACE

This book is intended as a text for a first course in the design and organization of digital computers. The book does not assume any prior knowledge of this subject from the reader. The text material has evolved from a course on the logical design of digital computers over a period of several years and was first used at the University of Ottawa and later at the University of Windsor. The book is suited for students of computer science or electrical engineering. The topics and the depth of presentations are based on the needs of practical design situations with which students can readily experiment through logic and in microprocessor laboratories.

Chapter 1 provides a general introduction to the structure and components of a digital computer, along with some notes on historical landmarks in the development of digital systems. Chapters 2 and 3 present the mathematical foundation and fundamentals of design procedures on combinational and sequential switching circuits. These two chapters contain numerous illustrative examples to provide practical knowledge and an understanding of important techniques and performance considerations. The first three chapters serve as introductory material to the remaining chapters and can be used for a one term course with some material selected from Chapters 4 and 5.

Chapter 4 discusses the design of computer components for arithmetic and logic operations, operation code decoding, addressing, accessing, and storing of information in memory. Basic storage elements used in the primary memory are also introduced. Chapter 5 discusses techniques for the organization, interconnection, and control of digital computer components constructed as functional modules. Illustrative examples in Chapters 4 and 5 provide adequate information to construct a laboratory prototype of a functionally complete digital computer by using commercially available integrated circuit chips and circuit boards. These two chapters along with selected material from Chapters 6, 7, and 8 can be used for a one term course.

Chapter 6 discusses various techniques in constructing memory modules and devices for primary and secondary storage requirements for a practical digital computer system. It presents the choices available to the user to meet his or her requirements. Chapter 7 discusses the details of arithmetic and logic circuits and compares the operation of hardwired control versus microprogrammed control now used increasingly in processor design.

Chapter 8 discusses techniques for input/output interfacing and the handling of interrupts to the processor.

Problems are given at the end of each chapter, and students will find these tremendously valuable.

We are indebted to many individuals for their assistance in the completion of this book. In particular, the students in the logical design course provided many valuable comments and suggestions.

S.I.A.
K.T.F.

CONTENTS

Chapter 1

INTRODUCTION TO DIGITAL COMPUTER STRUCTURE

1.1 EVOLUTIONARY DEVELOPMENTS

The history of problem solving devices (counters, calculators, and then computers) can be traced back to at least several hundred years B.C. The first known device, the Chinese Abacus, was in use in 600 B.C. This was an instrument for counting by transferring little bamboo rods from one group to another. Its descendent, Suanpan, is still used by the Chinese. Although these were simple enumeration (counting) devices, they proved adequate for performing complex calculations or computations via repetitive manual operations. At about 500 B.C., Egyptians are known to be using reckoning boards (counters, calculating devices) which used pebbles for enumeration. Greeks and Romans also used several forms of reckoning boards. These were called calculi (small stones).

About 800 A.D., Abu Jafar Mohammad ibm Musa al-Khowarizmi, a Persian scholar, published his work on the system of algebraic manipulation on numeric and symbolic expressions. Khowarizm today is the small Soviet city of Khiva. Al-Khowarizmi's work introduced manipulations based on the decimal number system, as opposed to the Roman numerals. In early 1200 Leonardo Fibonacci introduced to Latin Christendom the counting and algebraic manipulation of numbers and symbolic representations based on decimal numbers. Leonardo Fibonacci was born in Pisa, Italy and in adolescence joined his father, in Algeria, who was manager of a Pisan trade agency there. He received his education from a Muslim teacher and learned the use of calculating devices of that time. Despite Fibonacci's epoch-making work, the new method of calculation was long resisted by the merchants of Europe; many of them preferred to finger the abacus and write the results with Roman numerals.

About 1400 A.D., Persian astronomer Jamshid ben Masud Ghiath ed-Din al-Kashi developed many special purpose calculating devices which he used at the astronomical observatory in Samarkand. Many European mathematicians realized that the decimal number system opened the way to developments in mathematics (computation) which were impossible with the old letter numerals of Greeks, Romans, and Jews. In about 1600 A.D., the decimal number system finally replaced the Roman in England and America.

In 1642, Blaise Pascal, a French customs official, developed the first (decimal) adding machine using rotating wheels (gears). In 1671, von Leibniz of Germany extended the machine developed by Pascal to perform the operation of multiplication by adding extra gears (wheels). In 1823 Charles Babbage, a professor at Cambridge University, conceived the design of a differential engine (mechanization to perform all operations in a predetermined programmed sequence). The machine was to perform the instructions dictated by cards automatically, not stopping until an entire sequence of instructions had been completed. Babbage received a grant from the British Admiralty to develop this machine. However, it was not possible for him to realize the full complexity of this machine with the technology available during his time. Finally, in 1853, the successful implementation of this machine appeared in Sweden and the U.S.

In 1890, Herman Hollerith, a U.S. Census employee, conceived the idea of using holes punched in cards to store census data to be processed by machine. At this time keyboard machines were already quite common. The first computers in the U.S. were built using relays in 1940-42 at Harvard University in conjunction with IBM and Bell Laboratories. The first electronic (all vacuum tube) computer (ENIAC), was designed at the Moore School of Electrical Engineering, University of Pennsylvania in 1946. ENIAC lacked the flexibility of a general purpose computer. Perhaps the most significant conceptual development took place when John von Neumann proposed a general purpose stored program binary digital computer (EDVAC) as a successor to ENIAC mentioned previously. It was at this time that the present structure of digital computers—as shown in Fig. 1.6—was conceived and implemented. Since that time, high speed components and software developments have turned that computer into a powerful general purpose tool available to man for his problem solving needs, from very simple to highly complex ones. Soon, other more general computers such as EDVAC, SEAC, Whirlwind, Univac I, ORDVAC, and IBM 704 appeared on the market. Many of the earlier computers used magnetic drums for main memory. The coincident-current magnetic core storage invented in 1951 at MIT provided a much better form of random access main memory. In the mid-1950s computers with transistors in place of vacuum tubes first appeared. This led to the production of many commercially oriented computers. In the late 1950s developments in integrated circuit technology (packaging of an ever increasing number of transistors into a single electronic component) resulted in a new series of computers which were more powerful in operation, gave more performance per dollar, and offered a wider choice in configuration to meet user needs.

In the early and mid-1960s further improvements in integrated circuit technology resulted in the production of individual digital components, registers, memory, address, comparators, etc. as one integrated circuit. This resulted in the development of low-cost small computers known as mini computers. Integrated circuits were also produced for arithmetic and control functions of calculators. In the early 1970s large scale integration (LSI) technology made it possible to produce complex circuits requiring many thousands of transistors (gates) as a single electronic component known as an IC chip. General purpose processing units consisting of arithmetic, logic, and control circuitry all on a single chip called monolithic microprocessors were produced. At the same time, storage and input/output control requirements of the computer could also be satisfied by one or more chips. Small computers, known as micro computers, were designed by assembling a number of these chips on one or more circuit boards. Recent developments in technology to very large scale integration (VLSI) is opening up new possibilities in the design of computers and computer systems.

1.2 LOGICAL STRUCTURE

The digital computer is a general purpose device for solving problems. Many problems may also be solved manually or by a combination of manual operations and special purpose devices, although our ability to solve problems is severely limited in these cases. In order to understand the structure of digital computers, it is helpful to examine the requirements which must be satisfied before a problem can be solved. These requirements can be stated as follows:

1. We must know the possible sequences of operations to be performed in solving the problem. The sequence of operations actually performed in a given instance may vary depending on the initial values supplied to the problem solver and intermediate results generated during the solution. This sequence of operations is usually described in a form understandable to the problem solver.
2. We must provide the input data or information to be used in applying

the sequence of operations described in (i). Many different solutions or results can be obtained by providing different values of input data. For each instance of input data, a corresponding solution can be obtained without respecifying the procedure (consisting of a sequence of operations) to be used.

3. We must indicate what intermediate or final results are to be retained in storage for later recall and use, and what results are to be returned to the user as output, for each instance of the solution. This scheme for solving problems is illustrated in Fig. 1.1.

The device used to solve a problem must be capable of performing all possible operations specified in the problem-solving procedure. For example, if one of the operations is addition of two numbers then the device must know how to add, i.e., it must contain some component which is capable of performing addition. A digital computer usually contains components to perform operations such as addition, subtraction, multiplication, division, comparison of values to determine actual sequences of operations to be performed, and many more. Let us consider the simple operation of addition and how it can be realized in a digital computer. Suppose we want to add two numbers, called by names as X and Y. We may express the operation of addition as follows:

$$\text{Add } X, Y$$

or $$Z \leftarrow \text{SUM}(X, Y)$$

or $$Z \leftarrow X + Y \quad (1.1)$$

In the first case, the destination for result is implicit, i.e., it is placed (stored) in some predetermined location known to the problem solver. In the other two cases, the destination is explicitly called by name Z. The device for performing this operation can be represented very simply as

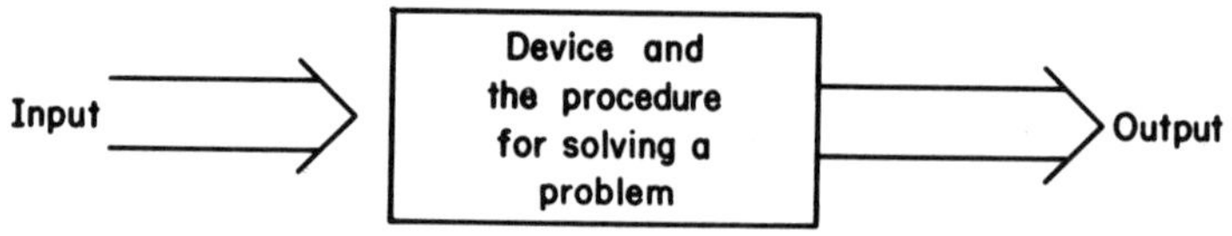

Figure 1.1 Problem solving involves application of a specified procedure to an instance of input information, producing an instance of output information

the block diagrams shown in Fig. 1.2. In the first realization (Fig. 1.2a) we have shown two input interfaces for the user to supply values of the two data items X,Y. However, input interfaces, such as keys, are expensive and it would be unwise and even impractical to provide separate interfaces for each possible data item at any given time. In the second realization (Fig. 1.2b), the user provides the value of the first data item (*X*) through the keys of the input terminal, then he presses a control key (shown as Register) to enable the transfer of input data into a storage area (shown as Register *A*). The user then provides the value of the second data item (*Y*) through the keys, and on pressing the control key (shown as Execute), the two numbers are added and the result appears on the output, which may be stored for use in other operations or simply displayed to the user. The registers serve the purpose of holding (retaining) input, output, or temporary data indefinitely. They may accept data at a slower rate and provide it later at a faster rate or vice versa. Thus they serve as buffers between their input and output interconnections.

The scheme of using a single input terminal is particularly appealing when many inputs are needed for a succession of operations, as we will discuss shortly. Suppose now we would like to construct a device which has components for performing two or more operations, e.g., addition (+), subtraction (−), multiplication (*), division (÷), etc. A block diagram for this device is shown in Fig. 1.3. In this case, after supplying the inputs as before, we can select the operation to be performed by pushing the appropriate key—add, subtract, multiply, or divide, etc., as desired. Pushing of a control key acts as a tap to open the flow of information to the appropriate component. Note that in Fig. 1.3., outputs of the adder, subtracter, multiplier, and divider are all connected to the output terminal. Again, control or enabling lines are used to ensure that the output from

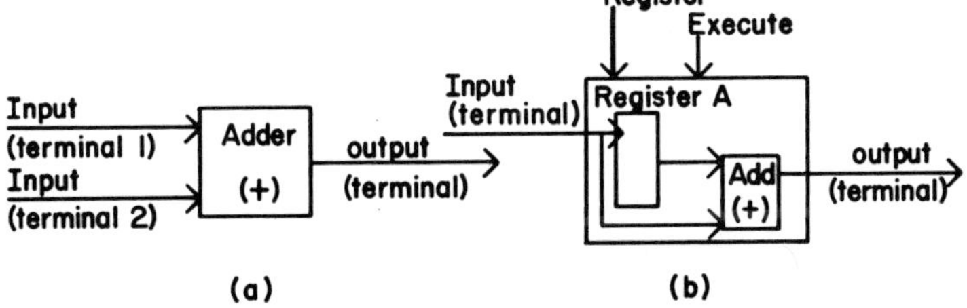

Figure 1.2 Realization of addition operation
(a) Two terminals for 2-inputs
(b) One terminal for two or more inputs

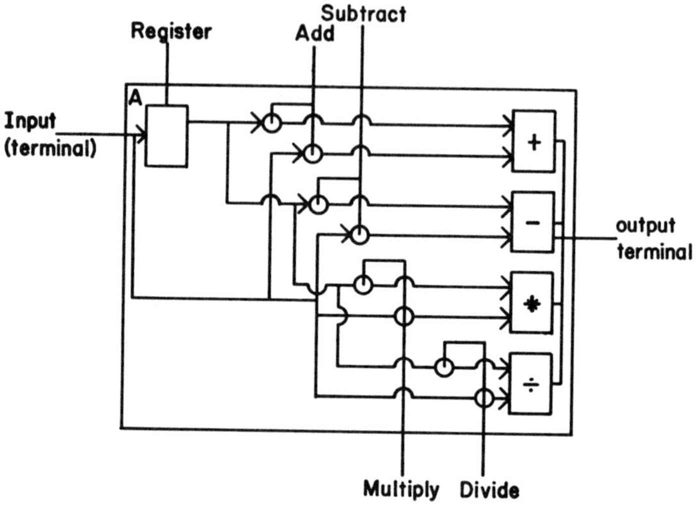

Figure 1.3 A device capable of selectively performing a specified operation

at most one of these is passed to the output terminal at any given time. Thus a sequence of control operations is necessary to perform a desired operation. Suppose we wish to perform subtraction. The following control sequence will take place.

1. Enable the input terminal and Register *A* to accept the value of a first data item, say *X*.
2. Enable the input terminal to accept the value of a second input data item, say *Y*.
3. Enable Register *A* and input terminal data to appear at the inputs of subtractor.
4. After enough time has elapsed so that the subtracter has completed the operation, it enables the subtracter's output to be routed to the output terminal or some register designated to receive the result of the operation just performed.

With the basic mechanism as described above properly operational, we can now solve any problem requiring addition, subtraction, multiplication or division. Suppose we want to evaluate the following expression:

$$(X - (X - Y)*Z) \div Y \qquad (1.2)$$

Let us assume that the result of evaluation is to be placed in some location, called S, i.e.,

$$S \leftarrow (X - (X - Y)*Z) \div Y \tag{1.3}$$

We may perform the following sequence of operations:

$$\begin{aligned} P &\leftarrow X - Y \\ Q &\leftarrow P * Z \\ R &\leftarrow X - Q \\ S &\leftarrow R \div Y \end{aligned} \tag{1.4}$$

As mentioned earlier, we can use registers to retain input values as well as temporary results. Also, we may want to solve the expression given in (1.3) or similar expressions for many different values of input data. Thus, it is desirable to construct a device which is also capable of storing the sequence of operations itself so that it can be repeatedly invoked by the user for different values of input data. The structure of this device is shown in Fig. 1.4.

It is possible, of course, to use fewer than 7 registers for the above sequence because some of the registers can be reused after they have served their purpose of supplying previously stored data for the intended

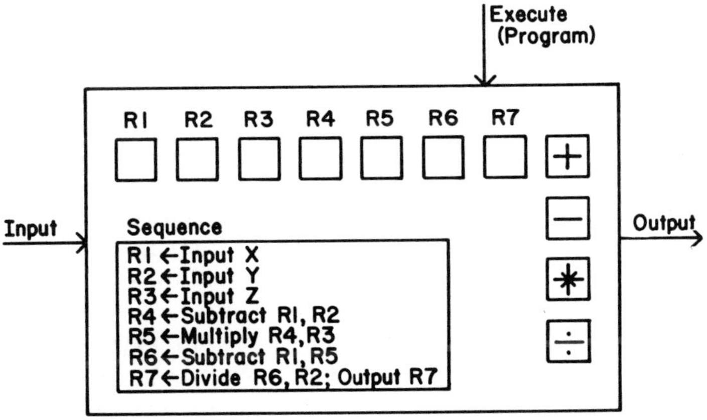

Figure 1.4 Device containing an arbitrary number of registers and space for storing the sequence of operations (program)

(program) in the available space and execute this program one or more times, for various instances of input data values producing corresponding instances of output. This device is known as a *Digital Computer*. It is automatic, i.e., it does not require manual intervention for executing each operation, step, or instruction of the program. Since the entire program consisting of a sequence of operations required to solve a problem is stored in the available space, the device is also known as a *stored program* Digital Computer.

As mentioned earlier, the sequence of operations to be executed by the computer is known as a program, and specifications of individual operations to be performed are called instructions (for the computer). The sequence shown in Fig. 1.4 is very simple, and each instruction of the sequence would be executed every time we press the Execute key. This is because this computer (device) does not have any components for making decisions based on examination of input data or comparison of other values and internal states of the device. This situation can be easily remedied by adding components for comparing values and branching to a certain instruction in the program. Now, only selected instructions of the program may be executed when the Execute key is pressed, using inputs supplied by the user.

Based on the preceding discussion, it now is possible to draw an overall logical structure of a digital computer as shown in Fig. 1.5. In a computer, all components are physically interconnected. However, flow of information to these components and their operation are controlled by a number of additional connections called control lines.

1.3 PHYSICAL COMPONENTS

In Section 1.2, we discussed the logical structure of digital computers and identified basic functions such as input, output, arithmetic and logic, storage and control. These functions are performed by suitably designed devices known as hardware components. Each component or unit can be designed separately and fabricated (assembled) using different parts to satisfy user needs in terms of cost, speed, and flexibility of operation. Later an appropriate choice of these devices can be made to configure a working digital computer. For example, the user can select the type of Input/Output (I/O) device which best satisfies his needs. Once a device has been selected, consideration must also be given to interfacing (connecting) the device to other physical components of the digital computer.

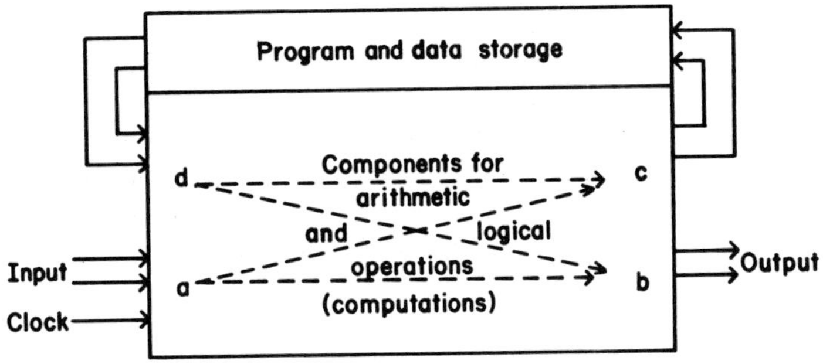

Figure 1.5 *Logical structure of a digital computer.* We see several types of possible computation:
(i) The input may simply be passed on to the output (path *ab*);
(ii) some computation may be performed on the input before storing it for subsequent recall (path *ac*)
(iii) previously stored information may be retrieved and transmitted to the output with or without computations (path *db*)
(iv) previously stored information may be retrieved, some computation on it, and restored (path *dc*). A clock is used as a control signal to operate components of the computer

In the case of an input device, connections must exist to enable the device and to route the data coming from the device to a register of the processing unit or to a location in memory. Connections must also exist to check whether the device is ready to send data, and appropriate timing signals must be provided to actually transfer the data from the device to a specified destination. The operation of the device depends on these hardware interconnections and execution of the program to initiate and terminate the transfer of data from a selected device to one of the many possible destinations. We will later examine the hardware interconnections and programs for I/O devices. For the moment, it will be sufficient to say that programs tailored to device characteristics are required to control its operation and to facilitate the transfer of information. These programs are called *driver routines* or simply *drivers*.

Similarly, the user can select a device for storing his programs and data. The first consideration in the choice of a storage device is the size or capacity of storage required and the basic unit of storage. We may express the capacity in units of words of storage. We then have to state the amount of information that one word can carry. Internally, the digital computer storage consists of on/off devices just like ordinary switches. Any device capable of representing only two possible states is known as a binary device, and its two states can be designated as 0 or 1. We can refer to this device as a storage cell which can hold one *bit* (abbreviated from *b*inary dig*it*) of information. A group or collection of these cells can be used to represent any information in encoded form. For example two binary cells can have four different designated states 00, 01, 10 and 11. These four states can be used to represent four different items. The ten possible values (0 to 9) of a decimal digit can be encoded by four binary cells and stored inside the computer as 0000, 0001, 0010, 0011, 0100, 0101, 0110, 0111, 1000 and 1001. Thus, we can express the storage requirements in terms of bits of information with the ability to address individual bits as units of information. However, for most storage requirements we prefer to move information in blocks of 8, 16, 32, etc., bits of information to or from the storage device. This scheme of addressing groups of bits economizes the number of address lines required to interface storage device to other physical components of the computer. Common units (blocks) of storage in small computers are 8-bits (1 *byte*) or 16-bits (2 *bytes*). The other important considerations in the selection of storage devices are speed of operations and whether the information stored stays there indefinitely (non-volatile) even if power supply to operate the device is turned off. The user must also consider whether he needs a storage device where the information stored once may never be changed (read only memory—ROM) or a device for storing different information from time to time. Thus we visualize storage as a distinct physical component of the computer because many different choices are possible.

The arithmetic and logic hardware of the computer (known as the ALU) is a physical component consisting of many parts internally connected together to perform functions described earlier. We can select an ALU with just a few arithmetic and logic functions and a few internal registers for holding temporary data or one with a large number of functions and internal registers depending on our requirements. The signals to route data in and out of the ALU, and initiate as well as terminate selected operations come from a device known as the Control Unit (CU). The CU is visualized as a separate device because it must provide an overall coordination of all devices in the system. However the interconnection between the ALU and CU are usually so extensive that we may often lump the two together as the *Central Processing Unit* (CPU) of the computer. We call it a central processing unit because those processing operations which are very specific to a device may be implemented within the device itself. The user can select an ALU and CU separately instead of an integrated CPU. In either case he makes this choice according to his specific needs for some desired speed of operation, and the kinds of other devices which he plans to use in the system. A block diagram of the digital computer, in terms of the physical components described earlier, is shown in Fig. 1.6.

The operation of a digital computer is, therefore, represented in terms of distinct components (modules) which receive, transform, and send information in accordance with the signals generated by the control unit.

Note also that the box representing the input, for example, in Fig. 1.6 has been labeled Input Unit rather than input device. This is because we may want to connect many input devices to the computer. Each one of these devices may have different interface hardware and control signal requirements. However, some of these requirements will be common to all input devices. We can lump these common requirements together into an Input Unit. Many input devices can interface to this unit through hardware individually specific to them. The Input Unit then establishes common interface with other components of the computer with standardized signals. The same is true for the Output Unit.

The Input Unit may receive information from one or more of the following input devices:

1. *Switches (for binary on/off, I/O input)*
   This is the simplest method (in terms of interfacing requirements) of providing input data to the computer because no conversion or

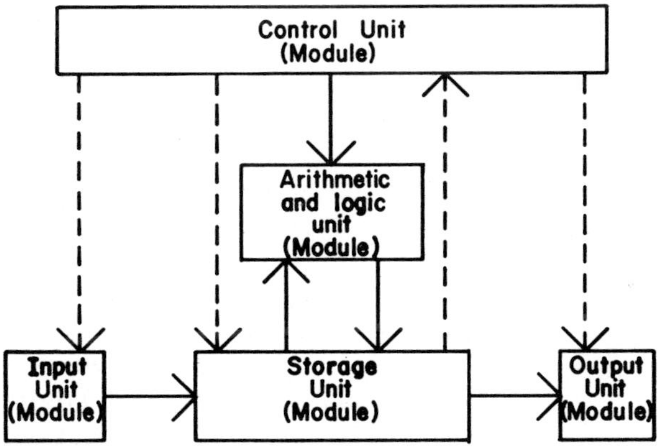

Figure 1.6 Physical components (modules) of a digital computer. The dotted lines represent control signals, and the solid lines represent flow of information.

operation. In the above case, we can store the sequence of operations decoding is required from what the device is sending and what the computer is designed to accept. The user sets the switches to on or off according to the input bit pattern, and then presses a control key (usually called ENTER) which tells the control unit that the device wants to send data; the control unit enables the device to put data on the data lines. From the data lines, the information is moved by the control unit either to a special internal register (usually called the *accumulator*) or to a location in storage previously identified by the user. Since hardware interconnections must exist from all the devices to the CPU, the only additional requirement is that the interface must ensure that CPU data lines remain unaffected by device connections. This is usually done by providing tri-state buffers between the input and CPU data lines. The operation of tri-state buffers is quite simple: in the absence of an enable signal from the CPU, they behave as if no connections exist between them and the CPU. When enabled, appropriate data received from the input device appears on the CPU data lines.

2. *Keyboard (for data consisting of alphabetic, numeric, and special characters)*

   Since there are many keys on the keyboard and each represents a different data or control character, they are all assigned separate binary codes within the hardware associated with the keyboard. These binary codes are usually composed of seven bits (called ASCII codes) and an additional bit may be added by the keyboard hardware for the purpose of error checking when the bit pattern moves from the device to the CPU data lines. To keep the interface between the device and Input (Interface) Unit simple, parallel data bits from the device are converted into serial bit patterns. The Input Unit receives these bits and assembles them in a buffer register of its own. When all data bits have been received, it puts them on the CPU data lines just as input from switches had appeared on these lines.

3. *Teletypewriter and Papertape Reader (for data consisting of alphabetic, numeric, and special characters)*

   The operation of a teletypewriter is similar to the keyboard operation. There are some differences in the interface hardware because of differences in timing and signal levels. Paper tape reader and punch devices are physically attached to the teletypewriter. Punching of paper tape allows preparation of input data off line or under computer control without having to provide it through the keyboard everytime.

4. *Card Reader*
   Input data may be punched on cards using a punching machine manually or under computer control. The cards usually have eighty columns and one ordinary character can be punched in each column. The holes are punched in one or more of the twelve positions in each column to represent the character in coded form (Hollerith code). The hardware associated with the device reads these 12-bit codes at a speed which is much higher than a paper tape reader and passes these bit patterns to the Input Unit, which may transmit them to the CPU data lines in two segments of eight bits. The entire code is then converted into appropriate internal code of the computer. Data can also be punched on the cards in other suitable forms which differ from single character per column. The computer program which reads these cards as input can perform appropriate translation into internal codes.

5. *Optical/Magnetic/Mark Sense Readers*
   The input data can also be provided in a variety of other forms, such as ordinary printed material (text) or bar codes, magnetized surfaces to represent message codes, or pencil marks in specified positions of printed cards. Appropriate devices are available to read this input and the user can select one or more of these to satisfy his requirements.

There are many other devices, such as magnetic tape, drum, disk, etc., which may be used as input devices for previously stored information. For the devices listed earlier, we may design a single input unit which is capable of receiving information from all of these devices, or we may design separate input units for different devices, and then another common input unit for receiving information from the device specific units. In the former case the design of a single input unit would be quite complex because unfortunately the input devices have developed historically with very few common features. In the latter case, we are attempting to distribute the operation of inputting data to device specific and device independant functions.

The output devices perform a reverse function, e.g., turn switches/

lamps on or off, print or video display, punch tape or cards, etc., all based on the intermediate or final results of program execution. The driver programs or routines perform conversions from internal form in computer memory or processor registers to codes acceptable by the output devices. The drivers also ensure proper timing and buffering requirements of the devices.

The storage or memory units (modules) usually consist of a large number of binary cells where each cell is capable of storing a one or a zero. The storage may be of semipermanent form, i.e., information stored once remains there unless modified or changed, thus permitting read-write operations. The storage may also be of permanent form, i.e., information once stored (through hardware programming) remains there and cannot be modified during normal operation. This form of memory is known as Read-Only-Memory (ROM) and it is very useful for programs to be used repeatedly without modification. The programs are accessible immediately after computer power is turned on and they are protected against inadvertent modification during computer operation.

Most of the main storage or memory is constructed so as to provide read-write access to a word (one, two, or more bytes) at random—known as random-access-memory (RAM). ROM also provides random access but we simply call it ROM rather RORAM. A combination of ROM and RAM provides what is generally referred to as the primary memory of the computer. All non-random access memory attached to a computer is known as auxiliary or secondary storage.

The two major technologies used for constructing memory units are magnetic and semiconductor. Within these two technologies there exist a variety of device fabrication techniques resulting in relatively slow to very high speed operation, small or high power consumption (dissipation), volatile or nonvolatile, static or dynamic, etc. We will study these devices in Chapter 4.

Arithmetic and logic units (modules) consist of circuits which perform addition, subtraction, comparison, etc. In order to perform these operations, input data in suitable form must be made available to these circuits. This input data is usually retrieved from memory and retained in registers provided for arithmetic and logic operations. Which operations are activated and in what particular sequence is based on activation/deactivation signals provided by the control unit. The control unit (module) derives these sequences by decoding the operation code bits of the program instructions. We will consider the design of arithmetic and logic unit as well as the control unit in Chapter 4 and Chapter 7, respectively.

1.4 INTERCONNECTION AND CONTROL

In Section 1.3, we briefly discussed the physical components or modules. In order to permit the flow of information among these modules and to control their operations, suitable interconnection of modules must be provided. For example, program instructions are read from an input device. This device may read information character-by-character and prepare it for transmission to the memory module. The memory module may have been designed to accept information one character at a time, two characters at a time, or more. Assume that the memory module accepts two characters at a time. Internally, each character may be coded by using eight bits. Therefore a 16-line interconnection is needed between the input unit transmission end and the memory unit receiving end. Since the input unit prepares the input it receives (i.e. suitable conversion or transformation of bit patterns), it requires one signal line to inform the memory unit that the data is ready. When the memory unit receives data, it must be informed of the location (address) in which this data must be stored. This is done by energizing the appropriate address lines. The bit pattern for energizing the address lines is provided by the control unit. Usually, the memory unit contains a data register (memory buffer register, MBR) for holding the data it receives, and an address register (memory address register, MAR) for holding the address of the location. Therefore, if the memory consists of 65536 (2^{16}) locations then a 16-bit address must be provided. This can be done by an 8-line interconnection which initially receives the first half of the address and loads it into the left half of the MAR and then receives the second half of the address and loads it into the right half of the MAR. Alternatively 16-line interconnection can be provided to transmit and receive all 16 bits of the address simultaneously. The memory unit must inform either the input or the control unit that it has (successfully) stored the information previously received, requiring another signal line. For output, the interconnection requirements are similar.

During most of the computation (execution of instructions), the flow of information and the associated interconnection requirements are as follows:

The control unit provides the address of the next instruction via the address lines. The instruction is retrieved and loaded into memory. The k operation code bits are passed onto the control unit decoder by k lines. The address bits are passed onto the arithmetic and logic unit for calculation of the actual address of the memory location containing data corresponding to the instruction now being decoded. The number of lines between the MBR of the memory unit and the arithmetic and logic unit is usually equal to the number of bits in the MBR. Thus the arithmetic and logic unit can send information to the MAR or send/receive information to/from the MBR via the same set of lines. A group of lines for transmitting data between devices is called a *bus*.

The control unit interconnections are also quite extensive because, in addition to sending signals to modules, it must frequently receive status information from the I/O units. The details will be discussed in Chapter 7.

1.5 USER INTERFACE

In general, the user need not be familiar with the detailed structure of the computer. He or she only needs to know those facilities of the structure which he can access to solve his problem. This, in turn, depends on the choice of the programming language. For example, in using a problem-oriented programming language such as FORTRAN or PL/I, the user view of the computer (interface) is limited to memory space which he or she can use by specifying arrays or named locations. These arrays or named locations (variables) may contain integers, floating point numbers, characters, etc.

To solve a problem on the computer, the user begins with problem specifications. These specifications are passed on to the programmer. The programmer may also be the one who first stated the problem specifications. In either case, the programmer develops details of the problem solution in an algorithmic or flowchart form. These details are then converted into a program using one of the programming languages. To run this program on the computer it may be punched on cards. Alternatively, we may sit at an interactive terminal (connected to the computer) and supply program statements through the keyboard. In either case the user has to supply some additional information so that his program can be run on the computer. For example, if the program has been written in a programming language other than the machine language, then another program, called a compiler for this language, must be run to accept the user program and to translate it into machine codes (program) which can then be executed (run) on the computer hardware.

In most medium to large scale computer systems there may be many users submitting their programs for execution. In this case users are issued identification numbers to indicate whether or not they are authorized to run their programs. Limits on storage space required and run time may also be imposed on user programs to ensure equitable sharing of computer resources by the community of users and detection of program malfunction. If many Input/Output devices are connected to the computer, and the user has referred to them merely by symbolic names in his program, then information must also be provided to change these names into device codes.

Thus the user interface may require only very simple actions or commands to run a program or a fairly elaborate set of actions or commands.

In this chapter we have introduced the digital computer in terms of the basic operations that it performs and the components used in its design and organization. The most elementary component in a digital computer is an on/off switch known as an electronic gate. A number of these gates are suitably connected to perform the operations such as addition, subtraction, comparison, storage, control, etc. Chapter 2 will discuss basic principles and techniques used in designing these gates and other more complex circuits based on suitable interconnection of these gates.

PROBLEMS

1.1 Figure 1.3 shows a device capable of selectively performing one of the operations +, −, *, and ÷ by pressing one of the corresponding add, subtract, multiply and divide buttons. Add another Register *B* so that pressing of the Register button transfers the contents of *A* into *B*, and simultaneously *A* gets loaded with the current input.

(a) Modify this device to include the following additional operations:
(i) ↑ (Power), (ii) $\sqrt{}$ (square root),
(iii) ↔ (Exchange) (iv) Ln (natural logarithm)
Add an extra connection from the output to the input of register A so that A can either be loaded externally by pressing the register button or be loaded from the output by pressing the store button.

(b) List the sequence of manual operations required to evaluate the expression $X \div \sqrt{(P - Q)} \uparrow R + S$ using the extended device.
NOTE: Assume that unary operators requiring one operand only use the contents of register *A*.

1.2 Figure 1.4 shows a device containing an arbitrary number of registers and space for storing the sequence of operations. This device contains components for performing the operations +, −, *, and ÷.

(a) Extend this device to include operations
BrZ (Branch to a specified location in memory space if the register specified contains a zero value)
BrL (Branch to a specified location in memory space if the register specified contains a numeric value less than zero)

BrH (Branch to a specified location in memory space if the register specified contains a numeric value greater than zero)

(b) Write an instruction sequence for this extended device to evaluate the expression $S \leftarrow (X - (X - Y) * Z) \div Y$. Avoid using the multiplication operation when either one of the operands is zero and avoid using the division operation when the divisor is zero.

1.3 Figure 1.5 shows the structure of a digital computer containing paths *ab*, *ac*, *dc*, and *db*. By removing some of these paths, the functional capability of the computer is reduced.

(a) What happens if only path *ab* is retained?
(b) What happens if only path *ac* is retained?
(c) What happens if only paths *ab* and *ac* are retained?

1.4 Redraw Figure 1.6 so that all I/O devices are connected to a single I/O (Control) unit.

1.5 It has been mentioned that EDVAC was the first binary digital computer. Earlier computers were designed to operate directly on decimal numbers. What are the advantages/disadvantages of binary versus decimal machines?

REFERENCES

1. Durant, W. *The History of Civilization: Part IV (The Age of Faith).* New York: Simon and Schuster, 1950.

2. Goldstine, H. H. *The Computer-from Pascal to von Neumann.* Princeton: Princeton University Press, 1972.

3. Richards, R. K. *Electronic Digital Systems.* New York: John Wiley & Sons, 1966.

Chapter 2

COMBINATIONAL DESIGN PROCEDURES

In the logical design of digital systems, the first main topic to be taken up is usually the design of combinational circuits. This is because combinational circuits have a special characteristic which makes them more readily understood. This is the fact that the output of a combinational circuit at any moment is dependent only on its input at that same moment. This means that no memory of the past information is present in a combinational circuit (although this is true only under ideal conditions). In this chapter, a switching algebra, which is needed in the analysis and synthesis of logical systems, is first introduced, and then in subsequent sections, the various aspects in the design of combinational circuits are investigated. These aspects include the definition and physical realization of logic gates, the various methods of representation of switching functions, the techniques used in the simplification of switching expressions, the description of systematic design procedures, and the actual circuit realization of switching expressions.

2.1 SWITCHING ALGEBRA

Consider a set S consisting of the two elements 0 and 1. In order to define a switching algebra, the following definition is made.

An algebraic or binary operation on the set $S = \{0,1\}$ is defined to be a mapping from $S \times S$ to S.

Two algebraic operations on S, OR, and AND, denoted by the symbols $+$ and . respectively, are defined as in Tables 2.1 and 2.2. For convenience, $x \cdot y$, where x and y are variables that can take on the values of either 0 or 1, will be written as xy.

Note that the value of the OR function is equal to the maximum of its two operands and the value of the AND function is equal to the minimum of its two operands.

+		y	
		0	1
x	0	0	1
	1	1	1

Table 2.1 OR operation

.		y	
		0	1
x	0	0	0
	1	0	1

Table 2.2 AND operation

Having introduced an algebraic operation, it is now possible to define a switching algebra.

A switching algebra in an algebraic system consisting of the set $S = \{0, 1\}$ and two algebraic operations + and . in which the following postulates hold:

Postulate I. $x + y$ and $xy \in S$ for all $x,y \in S$.

Postulate II. The operations + and . are commutative, i.e.

$$x + y = y + x \tag{2.1}$$

$$xy = yx \tag{2.2}$$

for all $x,y \in S$.

Postulate III. The operations + and . are distributive, i.e.

$$x + (yz) = (x + y)(x + z) \tag{2.3}$$

$$x(y + z) = (xy) + (xz) \tag{2.4}$$

for all $x,y,z \in S$.

Postulate IV.

$$x + 0 = 0 + x = x \tag{2.5}$$

$$1(x) = (x)1 = x \tag{2.6}$$

for all $x \in S$

Postulate V. For each $x \in S$, there exists an element in S, to be denoted by x' and referred to as the complement of x, such that

$$x + x' = 1 \tag{2.7}$$

$$xx' = 0 \tag{2.8}$$

The switching algebra established by this set of postulates forms the mathematical basis for the area of switching theory in which a switching or logical variable can assume one of two possible values. Some of the important properties of a switching algebra which can be derived and proved directly from these will now be stated without proof.

1. Associativity: For all $x,y,z \in S$

$$(x + y) + z = x + (y + z) \tag{2.9}$$

$$(xy)z = x(yz) \tag{2.10}$$

2. Idempotency: For all $x \in S$

$$x + x = x \tag{2.11}$$

$$xx = x \tag{2.12}$$

3. For all $x \in S$

$$x + 1 = 1 \tag{2.13}$$

$$x0 = 0 \tag{2.14}$$

4. It can be seen that complementation, also known as the NOT operation, is given by

$$1' = 0 \tag{2.15}$$

$$0' = 1 \tag{2.16}$$

One interesting and useful property of a switching algebra is the symmetry between the + and . operations and the constants 0 and 1. Thus equations (2.1)–(2.16) can all be grouped together in pairs and when one of the equations in each pair is given, the other in the same pair can be obtained simply by interchanging the + and . operations and also the constants 0 and 1. This property is known as the *principle of duality*. It applies to all theorems and results related to this switching algebra.

An expression in which one or more switching variables are connected through the use of algebraic operations (+, · and ′) is known as a switching or Boolean expression. A switching expression has the value 1 or 0 for any given set of values assigned to the variables. For example, the expression $x' + yz$ is evaluated as $0' + 0 \cdot 1 = 1 + 0 = 1$, directly from the previous postulates and laws of the switching algebra for $x = 0$, $y = 0$ and $z = 1$. A variable appearing in either complemented or uncomplemented form is known as a *literal.* In this way a Boolean expression defines a mapping and hence a *switching function.*

A switching function in a set of switching variables expresses the correspondence between each combination of values of the switching variables and an element in $S = \{0,1\}$. A switching function can thus be conveniently represented in a tabular form known as a *truth table* in which the left column enumerates all of the 2^n possible combinations of value of the n switching variables and the right column lists the corresponding values of the function.

Example. A switching function of three variables and its truth are shown in Table 2.3.

Notice that a switching function has a unique truth table representation but that it can be represented by a number of distinct expressions. Thus one method to show whether two expressions are equivalent (i.e. represent the same switching function) or not is to establish their truth tables to see if they are identical or not.

Before a logical system can be realized, switching expressions have to be obtained from the functional description of the system and transformed into some convenient forms. The switching expressions are then used to derive the realization, i.e., the design of the actual circuits.

In certain logical systems, we may not care what values the switching

$$F(x,y,z) = x' + yz$$

x	y	z	$F(x,y,z)$
0	0	0	1
0	0	1	1
0	1	0	1
0	1	1	1
1	0	0	0
1	0	1	0
1	1	0	0
1	1	1	1

Table 2.3 Truth table for example

function assumes for certain combinations of the switching variables. This might occur, for example, if a certain combination could never arise in practice. In these cases, it is necessary to introduce an additional logical value, to be known as a don't-care value and denoted by X, to represent the case when the value of the function can assume either a 0 or 1. A switching function in which at least one don't-care condition appears is said to be *incompletely specified.*

In the next section, we shall discuss how the description of a logical system can be translated into a set of switching expressions which can then be used to facilitate physical realization of the system.

2.2 TRANSLATING SIMPLE DESCRIPTIONS INTO SWITCHING EXPRESSIONS

We will illustrate the translation of verbal description into switching expressions through the use of examples. Let us first look at a very simple design example in which a half-adder is to be designed. The half-adder takes in two inputs, x_1 and x_2, and produces two outputs, the sum s and the carry c, as shown in Fig. 2.1. The sum s is 1 whenever one and only one of x_1 and x_2 is 1 and c is 1 only when both x_1 and x_2 are 1s.

From the description, it is easy to see that a switching expression for c is simply given by

$$c(x_1,x_2) = x_1x_2$$

and that for s is given by

$$s(x_1,x_2) = x_1x_2' + x_1'x_2$$

In the case of s, each of the terms x_1x_2' and $x_1'x_2$ represents the condition

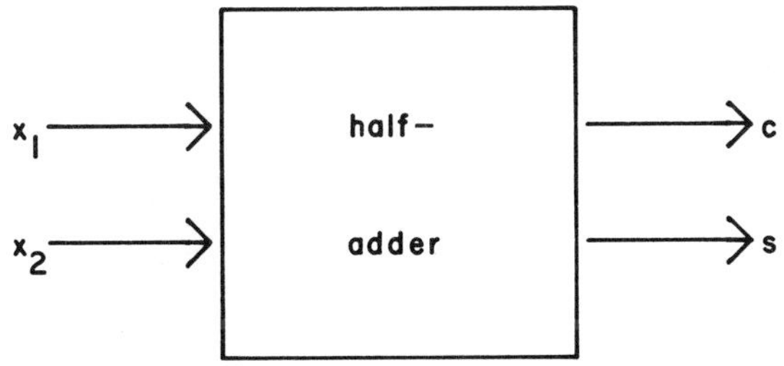

Figure 2.1 Block diagram of a half-adder

when one of x_1 and x_2 assumes a 1 and the other assumes an 0. The function describing s is also known as an exclusive-or operation ($x_1 \oplus x_2$) with symbol XOR, as shown in Table 2.4.

	$\oplus$	x_2: 0	1
x_1	0	0	1
	1	1	0

Table 2.4 XOR operation

In the previous example, the switching expressions can readily be obtained by inspection from the very basic description of the logical system to be designed. This, of course, is not generally possible, especially in more complex systems. However, a well analyzed and orderly description will very often enable the corresponding expressions to be readily derived for many complex systems.

Suppose we want to design a traffic light controller to regulate the traffic through an intersection of two one-way streets as shown in Fig. 2.2. Let

T_A be a counter which counts down, from a preset initial value, to 0 in 20 seconds, at which time a signal, denoted by t_A, will be produced.

T_B be a counter which counts down, from a preset initial value, to 0 in 60 seconds, at which time a signal, denoted by t_B will be produced.

T_C be a counter which is used to count the number of cars approaching the intersection on street A; T_C is initially set at 0.

A signal denoted by c will be produced by counter T_C once it reaches a value of 5. Signals S_A, S_B and S_C will be used to reset the counters T_A, T_B and T_C respectively to their appropriate initial values.

Assume that the following operational requirements are to be satisfied for the controller:

1. A green light for street A is to be turned on whenever the green light for street B has been on for a period exceeding 60 seconds or whenever there are 5 or more cars waiting to cross the intersection on street A.

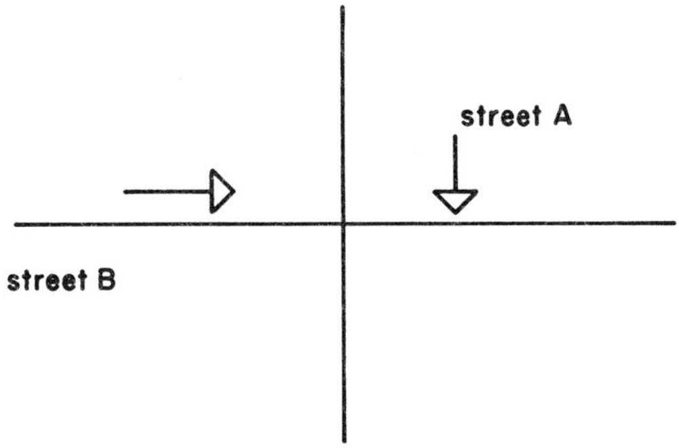

Figure 2.2 Traffic intersection

2. A green light for street B is to be turned on whenever the green light for street A has been on for a period exceeding 20 seconds, and there are less than 5 cars waiting on street A to cross the intersection.

3. For simplicity, the presence of an amber light for both streets will be ignored so that turning on a green light for street A will necessarily mean turning on a red light for street B and vice versa.

Let us use the switching variables l_A and l_B to represent signals for turning on a green light for street A and for street B respectively. Thus l_A and l_B also represent signals for turning on a red light for street B and for street A respectively. Then, by inspection we can obtain the following expressions for l_A and l_B:

$$l_A = c + t_B$$

$$l_B = t_A c'$$

The signal l_A is also used to reset counters T_A and T_C to their initial values and the signal l_B is also used to reset counter T_B to its initial value:

$$S_A = c + t_B$$

$$S_B = t_A c'$$

$$S_C = c + t_B$$

The required switching functions have thus been obtained from the original description of the system without much difficulty. This is possible mostly because the operational requirements of the system have been adequately analyzed and the description presented in an orderly manner.

In the example, the logical structures of the counters used are not indicated; the counters are simply assumed to be capable of producing signals whenever certain conditions are satisfied. Counters are not entirely combinational circuit components and will be described in more detail in a later chapter when sequential circuits are discussed. The entire design of the simplified traffic light controller will also be completed at that time when all the essential tools have been developed.

2.3 LOGIC GATES AND IMPLEMENTATION OF SWITCHING EXPRESSIONS

The algebraic operations AND, OR, NOT, and XOR are implemented through logic gates. The four types of logic gates are represented by the symbols shown in Fig. 2.3. There is a variety of technologies which can be used to realize a logic gate. In addition, it is necessary to distinguish between the use of positive logic, in which a high voltage represents a

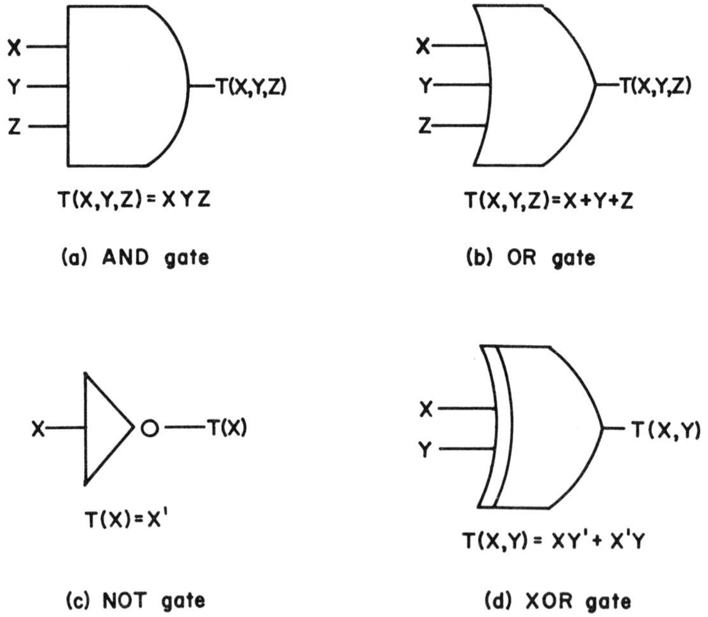

Figure 2.3 Gate symbols

logical 1 and a low voltage represents a logical 0, and the use of negative logic, in which a high voltage represents a logical 0 and a low voltage represents a logical 1, or an appropriate combination of the positive or negative logic at the inputs or outputs.

As an example, Fig. 2.4 shows two diode circuits which can be used as both AND gates and OR gates depending on the type of logic used. The circuit in Fig. 2.4(a) is an AND gate for positive logic and an OR gate for negative logic. On the other hand, the circuit in Fig. 2.4(b) is an AND gate for negative logic and an OR gate for positive logic.

Metal-oxide-semiconductor (MOS) circuits are becoming increasingly popular because they lend themselves readily to medium and large scale integration (MSI and LSI) circuit productions. Fig. 2.5 shows two MOS circuits which can be used to realize the AND gate, and the OR gate. MOS circuits have the additional advantage of requiring relatively small power consumption.

The realization of a NOT gate generally requires the use of a transistor for inversion of the logical level. Fig. 2.6(a) shows a bipolar transistor inverter circuit and Fig. 2.6(b) shows an FET (field-effect transistor) inverter circuit.

Due to the inverting property of transistors, circuits built from transistors tend to exhibit NOT-AND and NOT-OR behaviours. Therefore it is convenient to introduce two additional types of gates to take care of the NOT-AND and NOT-OR operations and these two types of gates are known as NAND and NOR gates respectively. The symbols used to rep-

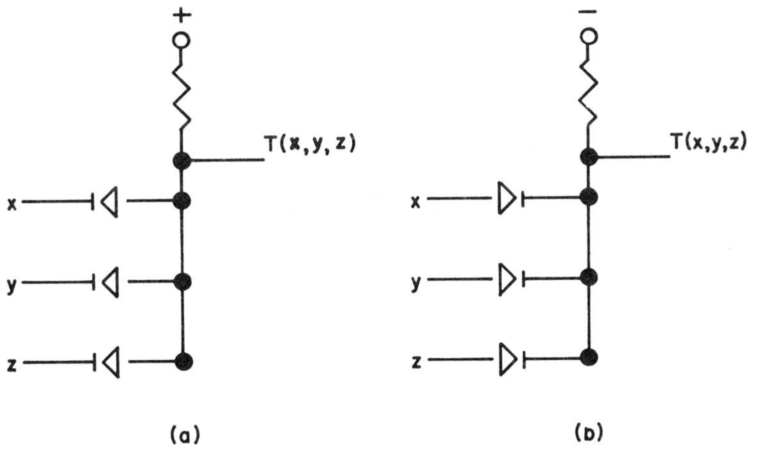

Figure 2.4 Diode circuits

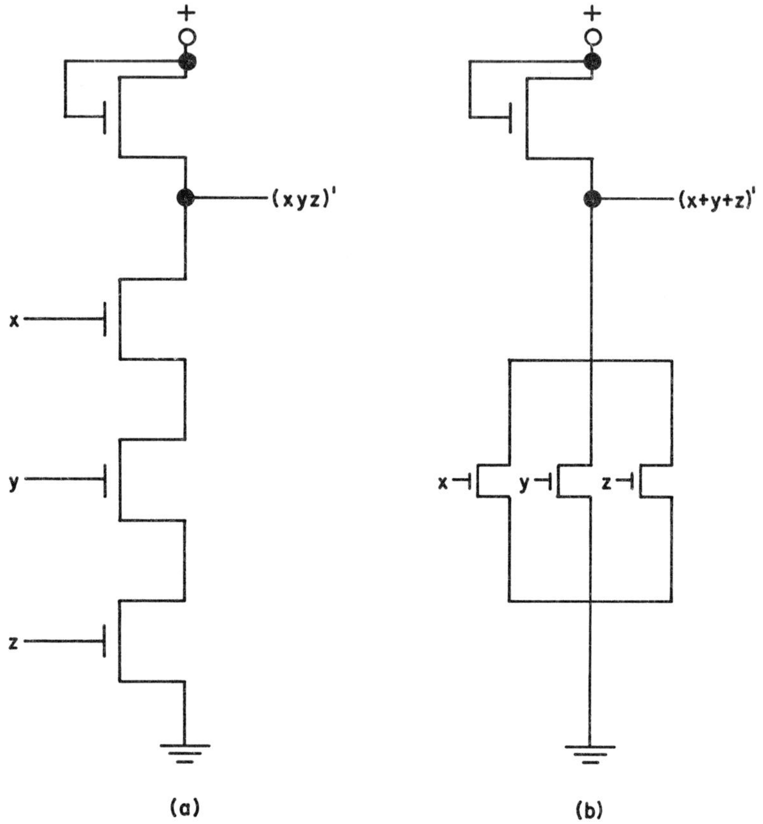

Figure 2.5 MOS circuits

resent these two gates are shown in Fig. 2.7. NAND and NOR gates are especially important when dealing with integrated circuits where the devices are predominately transistors. Fig. 2.8 shows one possible implementation of the switching function

$$T(x,y,z) = x'y + z$$

using only NAND gates. In a later section, a systematic procedure for implementing switching functions using NAND or NOR gates will be introduced. Note that single input NAND and NOR gates are logically equivalent to NOT gates.

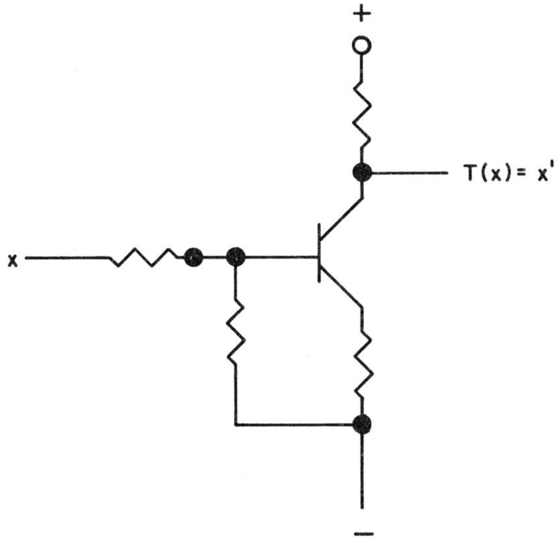

(a)

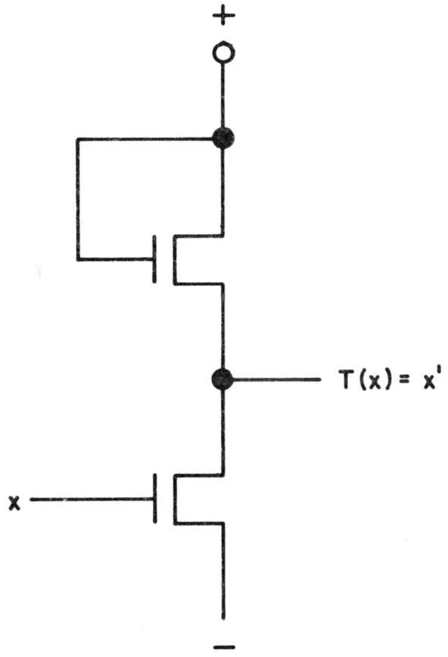

(b)

Figure 2.6 Inverter circuits

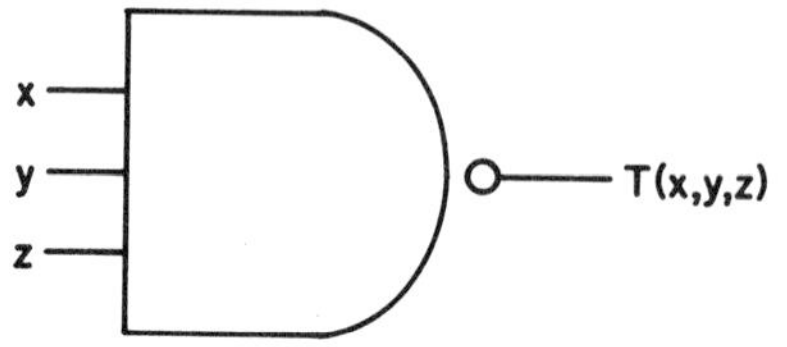

$T(x,y,z) = (xyz)'$

x	y	z	$T(x,y,z)$
0	0	0	1
0	0	1	1
0	1	0	1
0	1	1	1
1	0	0	1
1	0	1	1
1	1	0	1
1	1	1	0

(a) NAND gate

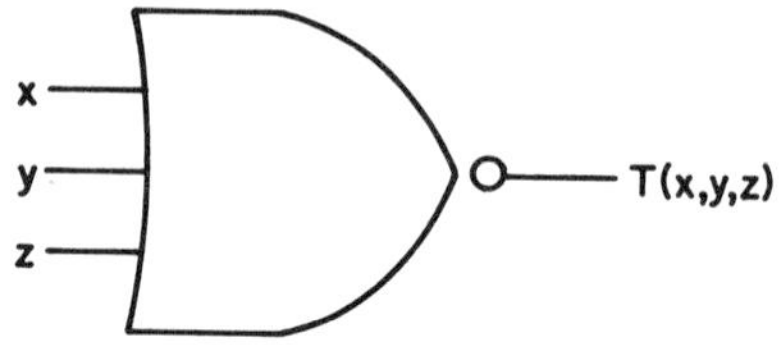

$T(x,y,z) = (x + y + z)'$

x	y	z	$T(x,y,z)$
0	0	0	1
0	0	1	0
0	1	0	0
0	1	1	0
1	0	0	0
1	0	1	0
1	1	0	0
1	1	1	0

(b) NOR gate

Figure 2.7 NAND and NOR gates

A MOS circuit realizing the same switching function as that shown in Fig. 2.8 is shown in Fig. 2.9.

It can be shown that any switching function can be implemented using any one and only one of the following sets of logic gates.

$$\{NAND\}, \{NOR\}, \{NOT,OR\}, \{NOT,AND\}$$

A set of gates which is sufficient to realize any switching functions is said to be *logically complete*. Note that {AND} is not a logically complete set of gates.

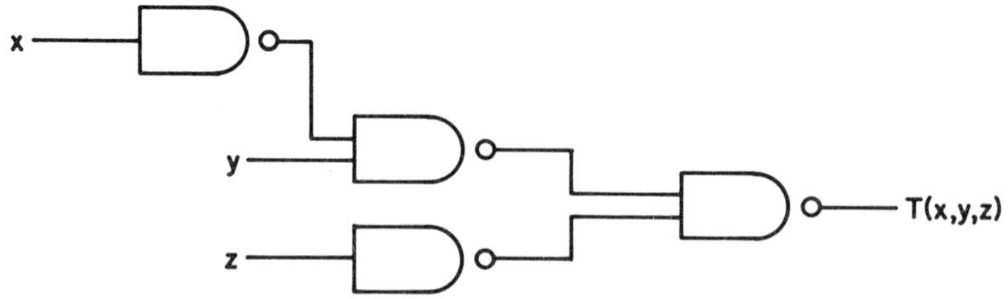

Figure 2.8 NAND gate implementation of $T(x, y, z) = x'y + z$

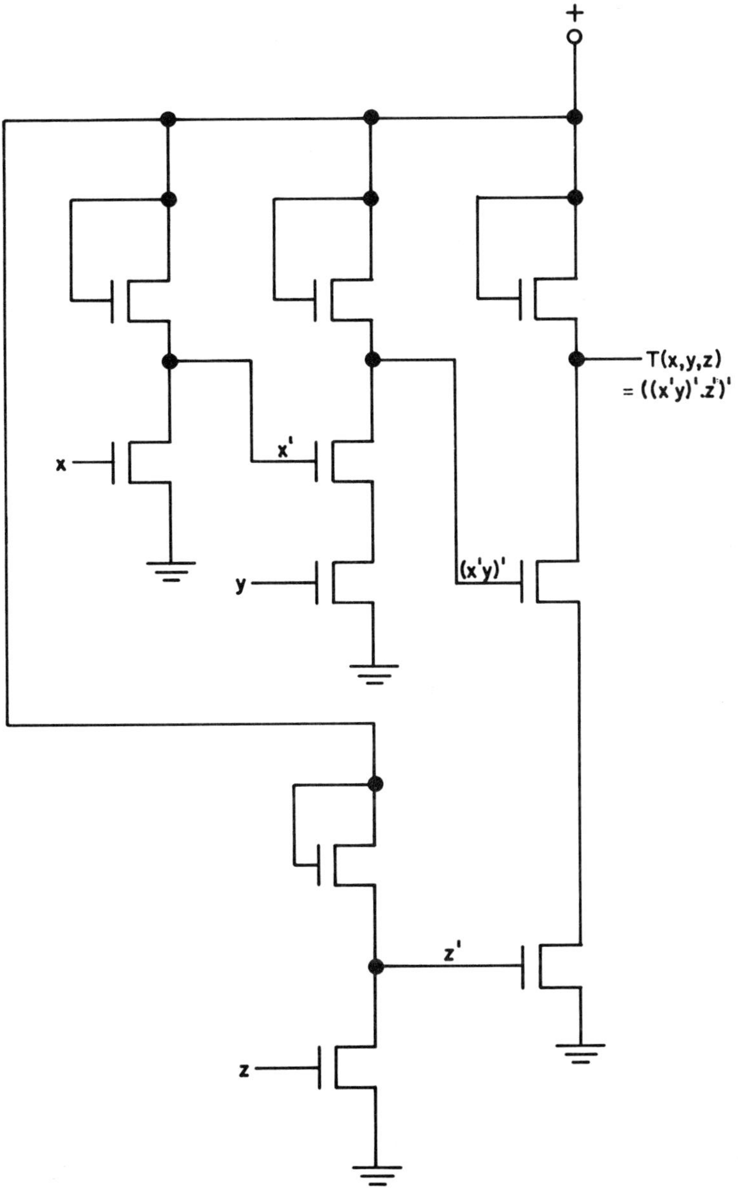

Figure 2.9 MOS circuit realization of the function $T(x,y,z) = x'y + z$

It has to be noted that although it is theoretically possible to design various circuits without much knowledge of the actual hardware involved by assuming ideal realization conditions, the actual characteristics of various hardware devices generally place significant limitations on the flexibility and feasibility of a design. For example, in a MOS gate, the number of field-effect transistors (FETs) that can be placed in parallel or in series is not unlimited and the number of gates that can be placed in cascade without some form of additional driver circuits is also not unlimited. Hence, a design may be logically correct but may not be realizable by one-to-one correspondence of the logical-versus-physical components.

2.4 MANIPULATION OF SWITCHING EXPRESSIONS

Two different switching expressions may represent the same switching function. It is then natural to suspect that it will very often be advantageous to choose the simplest expression to represent a switching function. This is indeed generally valid since a simpler expression will usually require for realization a circuit of less complexity but the term "simpler" requires the choice of a cost criterion which we will discuss in a later section. So for the moment the term "simpler" will be loosely used. This section deals with the manipulation of switching expressions.

A number of properties of the switching algebra provide the basic tools for the simplification of switching expressions. In addition to the previous postulates these include the following:

Let $x,y \in S = \{0,1\}$.

1.
$$(x')' = x \tag{2.17}$$

2.
$$x + xy = x \tag{2.18}$$

$$x(x + y) = x \tag{2.19}$$

3. (De Morgan's Theorem)

$$(x + y)' = x'y' \tag{2.20}$$

$$(xy)' = x' + y' \tag{2.21}$$

4.
$$x + x'y = x + y \tag{2.22}$$

$$x(x' + y) = xy \tag{2.23}$$

All these properties can be proved from the postulates in Section 2.1. For example, for (2.22),

$$\begin{aligned} x + x'y = x\cdot 1 + x'y &= x\cdot(1 + y) + x'y \\ &= x + xy + x'y \\ &= x + (x + x')\cdot y \\ &= x + y \end{aligned}$$

Example: Simplify the expression $xy'z + x'z + yz$.

$$\begin{aligned} xy'z + x'z + yz &= z(xy' + x' + y) && \text{by distributivity} \\ &= z(y' + x' + y) && \text{by (2.22)} \\ &= z(1 + x') && \text{by Postulate V} \\ &= z && \text{by (2.13)} \end{aligned}$$

Simplification of switching expressions is essential to the minimization of the complexity of circuits realizing switching functions. Additional tools will be developed in a later section to obtain simple switching expressions for switching functions. At this moment let us go into more detail on the methods of the representation of switching function.

There are two special forms of expressions representing switching functions which are of particular interest. Expressions of these forms are said to be in canonical or standard forms. An expression of the first canonical form is known as a canonical *sum of products* or a disjunctive normal expression. It is formed through the summation of a number of product terms where each product term contains each and every one of the switching variables as factors in either complemented or uncomplemented form. Such a product term is known as a *minterm*. Each of the minterms represents one particular combination of the switching variables for which the value of the switching function is 1.

Example: The canonical sum of products for the function $T(x,y,z) = xy'z + z + yz$ is the sum of four minterms

$$x'y'z + x'yz + xy'z + xyz$$

It is often convenient to represent each of the minterms by the decimal code corresponding to the combination of the variables making up the minterm. Thus, the minterm $x'y'z$ can be represented by the code 1, corresponding to the combination 001. Thus the function in the above example can be given as

$$T(x,y,z) = \Sigma(1,3,5,7)$$

Note that this method of representation requires a consistent ordering of the switching variables. Thus in the above expression, the variable x is regarded as the most significant binary place. We will use the term vector to denote an ordered combination of switching variables.

The second canonical form is known as a canonical *product of sums* or a conjunctive normal form. It is formed by multiplying a number of sum terms together where each sum term, also known as a *maxterm,* is formed by summing up all possible switching variables in complemented or uncomplemented form. Each of the maxterms stands for one particular combination for which the switching function has a value of logical 0 and can analogously be represented by a decimal code. For example, the maxterm $x + y + z$ evaluates to 0 for the particular combination of $x = 0$, $y = 0$ and $z = 0$ and thus can be represented by the code 0.

Example: The canonical product of sums for the function

$T(x,y,z) = xy'z + x' + yz$ is

$$(x + y + z)(x + y' + z)(x' + y' + z)(x' + y + z)$$

and representation in decimal codes is

$$T(x,y,z) = \Pi(0,2,4,6)$$

indicating a value of 0 whenever one of the vectors with decimal codes of 0, 2, 4 and 6 occurs.

The two canonical forms of a switching function can readily be obtained through the use of a result given in the following theorem:

Theorem 2.1–(Shannon's Expansion Theorem) [4,5,7,9] Any switching function $T(x_1,x_2,\ldots,x_n)$ can be expanded into one of the following two forms:

$$T(x_1,x_2,\ldots,x_n) = x_1T(1,x_2,\ldots,x_n) + x_1'T(0,x_2,\ldots,x_n) \tag{2.24}$$

$$T(x_1,x_2,\ldots,x_n) = (x_1 + T(0,x_2,\ldots,x_n))(x_1' + T(1,x_2,\ldots,x_n)) \tag{2.25}$$

where $T(1,x_2,\ldots,x_n)$ and $T(0,x_2,\ldots,x_n)$ are the function T evaluated at $(1,x_2,\ldots,x_n)$ and $(0,x_2,\ldots,x_n)$ respectively.

Repeated applications of one of the expansion equations (2.24) and (2.25) for each of the switching variables on a switching function result in an expression in one of the canonical forms.

Example:

$$\begin{aligned} T(x,y,z) &= xy'z + x'z + yz \\ &= x(y'z + yz) + x'(z + yz) \\ &= xy'z + xyz + x'z \\ &= y(xz + x'z) + y'(xz + x'z) \\ &= xyz + x'yz + xy'z + x'y'z \end{aligned}$$

Note that expansion around the third variable z is superfluous in this case since all product terms already have z literals. However in general an n-variable expression will require n expansions.

It can easily be seen that the two canonical forms are duals. The complement of one form can be obtained from the other one by simply interchanging the two operations + and . and a literal by its complement. Since the disjunctive normal form is a more natural form of representation and is easier to manipulate, we will limit our attention to this form of representation from this point onward unless otherwise indicated.

An expression given in the form of a summation of product terms (not necessarily minterms) can most readily be implemented by a two level AND-OR circuit in which the output of each of the AND gates acts as an input to the single OR gate. If only uncomplemented primary, i.e., switching variable inputs are available, then another level of gates, inverter gates, may be needed. An AND-OR circuit implementing the function

$$T(x,y,z) = xy'z + x'z + yz$$

is shown in Fig. 2.10(a), and an AND-OR circuit with inverter gates implementing the same function is shown in Fig. 2.10(b). In the next section we will show how simpler realizations can be derived from this simple canonical expression.

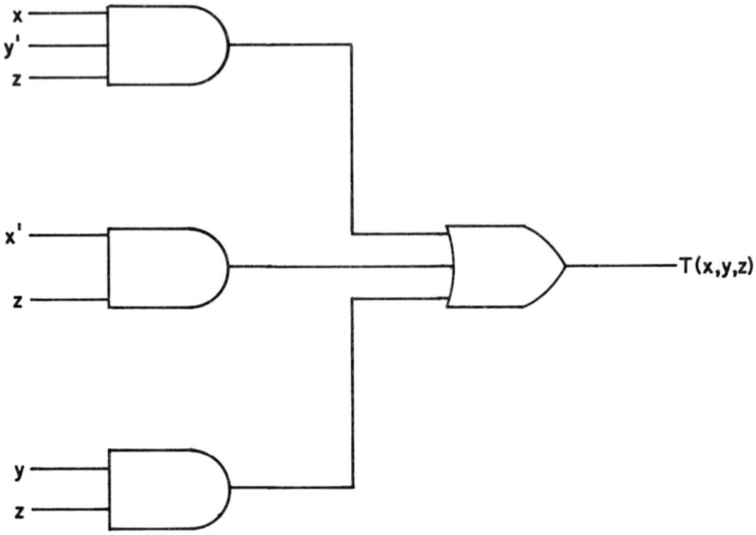

(a)

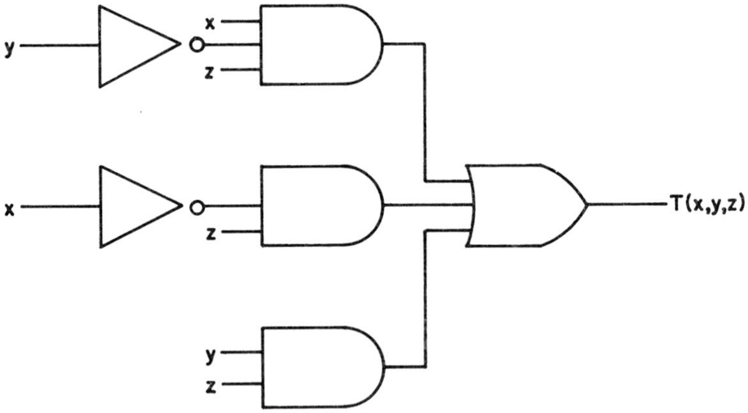

(b)

Figure 2.10 AND-OR circuits

2.5 LOGIC DESIGN PROCEDURES AND PERFORMANCE CONSIDERATIONS

It is possible to represent a given switching function by a number of expressions. In this section, systematic procedures will be described to obtain an expression for a switching function, which is minimal with respect to some prechosen cost criteria.

There are several cost criteria that are often used and they are the following:

1. Literal cost criterion according to which a minimal expression is one in which the number of literals is minimum.
2. Input cost criterion according to which a minimal expression is one in which the number of inputs, all assumed to be in uncomplemented form, is minimum.
3. Gate cost criterion according to which a minimal expression is the sum of product form expression with the minimum number of terms.
4. Gate-literal cost criterion is almost the same as the gate cost criterion defined in (iii) except that now, the minimal term sum of products expression with the minimum number of literals is chosen.
5. Chip cost criterion according to which a minimal expression is one that utilizes the least number of standard IC chips for implementation.
6. Interconnection cost criterion according to which a minimal expression is one that utilizes the least number of interconnection lines for implementation with standard IC chips.

Treatment of the criteria in (v) and (vi) requires a discussion on some very advanced implementation techniques and the reader is referred to other sources for these topics [1,6].

For simplicity, only the gate-literal cost criterion in (iv) will be considered unless otherwise indicated.

The first procedure is based on the use of Karnaugh maps. A Karnaugh map is actually just one form of representation of a switching function truth table in which the presence or absence of a minterm in the switching function is indicated by an entry of either a 1 or a 0 respectively in a table-like map. Fig. 2.11 is a Karnaugh map for the function $T(x_1,x_2,x_3,x_4) = \Sigma(2,5,6,7,10,14)$. Note the peculiar order in which the squares are labelled.

In a Karnaugh map, two squares or 0-cubes as they are called are adjacent to each other only when, in the two corresponding combinations

x_1x_2 \ x_3x_4	00	01	11	10
00	0	0	0	1
01	0	1	1	1
11	0	0	0	1
10	0	0	0	1

Figure 2.11 Example of Karnaugh map

of the switching variables, exactly one of the variables assume different logical values. For example, the two 0-cubes corresponding to minterms $x_1'x_2x_3'x_4$ (0101) and $x_1'x_2x_3x_4$ (0111) are adjacent to each and so are those corresponding to minterms $x_1'x_2'x_3'x_4$ (0001) and $x_1x_2'x_3'x_4$ (1001):

x_3x_4

x_1x_2				
		0001		
		0101	0111	
		1001		

Notice that the top and bottom rows are adjacent as are the left and right columns. This is due to the planar representation of a 4-dimensional space. Thus two 0-cubes having identical entries can be combined to form a 1-cube which corresponds to a term in which one of the variables is omitted. This combining process can be continued further for 1-cube, 2-cubes and so on to obtain cubes corresponding to terms with fewer and fewer literals. Thus, it can easily be seen that the gate-literal cost criterion is directly related to the identification of the smallest number of cubes of maximum sizes which together implement the switching function. The smaller the number of cubes required, the smaller is the number of gates required for realization. Maximizing the sizes of the cubes minimizes the number of literals involved.

We can thus formulate the following procedure for realizing a switching function subject to the gate-literal criterion using a two-level AND-OR circuit

Step 1: A Karnaugh map for the switching function is drawn.

Step 2: Maximum-size cubes are formed for squares with 1-entries.

Step 3: The smallest number of cubes is identified to implement the given switching function.

Step 4: The identified cubes are used to obtain an AND-OR circuit.

Example: The function represented by the Karnaugh map in Fig. 2.11 can be implemented by the following expression made up of two product terms:

$$x_3 x_4' + x_1' x_2 x_4$$

The cubes corresponding to the two terms are shown in Fig. 2.12 (inside brackets), and the AND-OR realization is shown in Fig. 2.13.

The above procedure can be extended to include don't-care conditions. A 0-cube with a don't-care entry can be combined with an 0-cube with a 1 entry to produce a 1-cube if necessary; this simply means that a 1 has been assigned to the don't-care entry. This is straightforward and will not be pursued further except for the following example.

Example: The AND-OR circuit realizing the function shown in the Karnaugh map in Fig. 2.14 is shown in Fig. 2.15.

The second procedure (Quine-McCluskey Method) is based on essen-

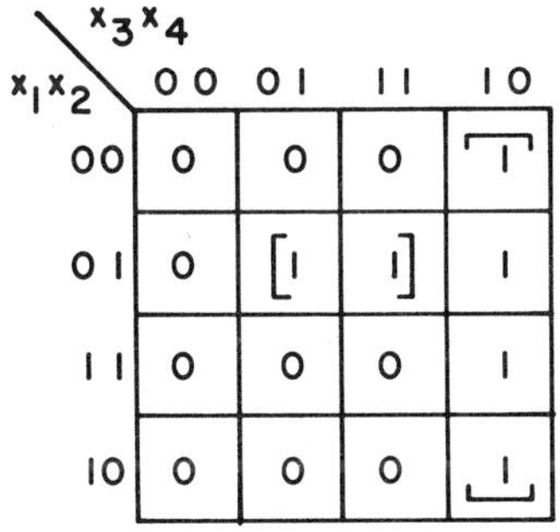

Figure 2.12 Karnaugh map for example

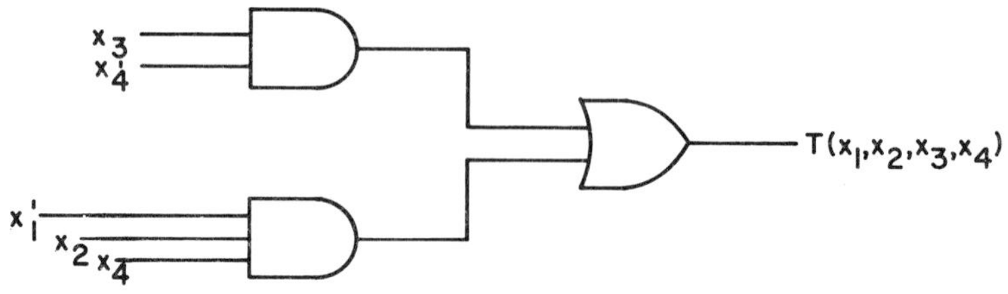

Figure 2.13 Realization of example

tially the same principles as the first one, except that it is a tabulation method rather than a mapping method. The mapping method can conveniently be applied to functions of at most five variables and is useful for manual computation since humans are adept at visualizing pattern. In addition, the processes of identifying and combining adjacent cubes and of the subsequent choosing of the cubes to implement the function are accomplished on a trial and error basis. On the other hand, the tabulation method offers a more systematic way to obtain the minimal expression and can easily be programmed to run on a computer, although such a program has run-time and storage requirements that are very data-sensitive and is generally limited to 15–20-variable problems. It should be pointed out that for bigger problems the two-level realization would probably be unfeasibly expensive.

This second procedure will be introduced through an example. Consider the following function

$$T(x_1,x_2,x_3,x_4) = \Sigma(2,5,6,7,10,14)$$

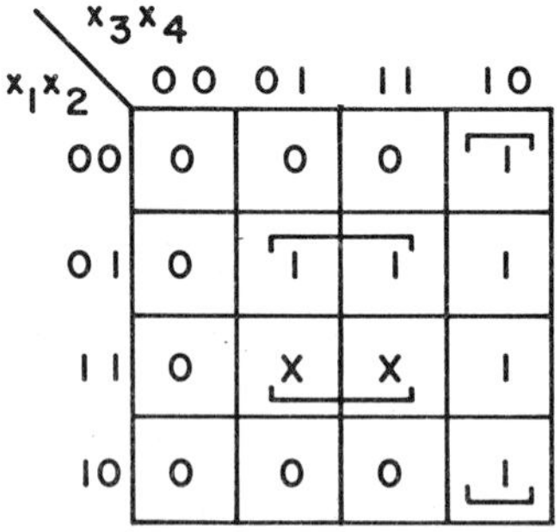

Figure 2.14 Karnaugh map for example

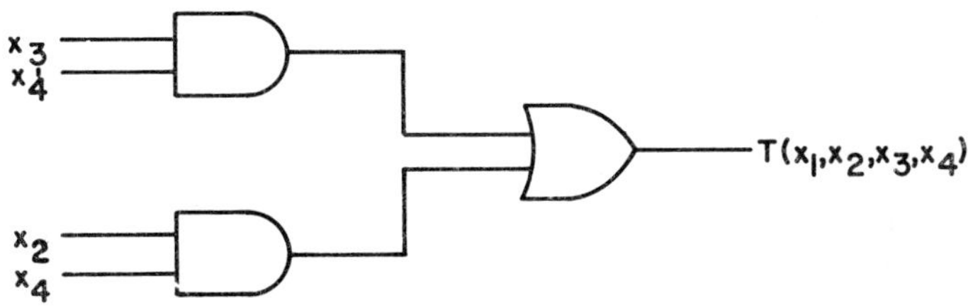

Figure 2.15 Realization of example

The first step consists of arranging the minterms of the function in groups Si as in Fig. 2.16(a) where group Si has minterms with i ones. Minterms are grouped according to the number of 1's in the switching variables. A group of minterms with fewer 1s is placed on top of one with more 1s. In the example, if 0000 is also a minterm, then there will be one more group at the very top. It can easily be seen that only minterms in adjacent groups can be combined together and this fact is made use of in the next step.

The second step is to obtain all possible combinations of pairs of min-

		x_1	x_2	x_3	x_4	
S1	2:	0	0	1	0	✓
S2	5:	0	1	0	1	✓
	6:	0	1	1	0	✓
	10:	1	0	1	0	✓
S3	7:	0	1	1	1	✓
	14:	1	1	1	0	✓

(a)

	x_1	x_2	x_3	x_4	
2,6:	0	—	1	0	✓
2,10:	—	0	1	0	✓
5,7:	0	1	—	1	
6,7:	0	1	1	—	
6,14:	—	1	1	0	✓
10,14:	1	—	1	0	✓

(b)

	x_1	x_2	x_3	x_4
2,6,10,14:	—	—	1	0

(c)

	2	5	6	7	10	14
x_3x_4'	x		x		x	x
$x_1'x_2x_3$			x	x		
$x_1'x_2x_4$		x		x		

(d)

Figure 2.16 Tabulation method

terms to form another set of groups of terms as shown in Fig. 2.16(b). Note that checkmarks are made on those minterms in Fig. 2.16(a) which have been successfully combined while a dash is made in Fig. 2.16(b) to indicate the particular variable involved in the combination.

Using the new set of groups of terms, the combining process as described in the second step is repeated successively until we arrive at a new set of groups of terms for which further combination is impossible. In the example, the final group is shown in Fig. 2.16(c).

Those terms which have not been checkmarked are picked out and these are known as the *prime implicants*. Formally a prime implicant is a term of a function which is covered by the function such that the deletion of any one-literal from the term results in a new term which is not covered by the function. A term is covered by a function if whenever the term assumes a value of 1, then the function assumes a value of 1. It is evident that the cubes of maximum sizes used to implement a function in the mapping method actually correspond to prime implicants.

The next task is to form a prime implicant chart (covering table) showing the covering of the minterms by the various prime implicants. This is done in Fig. 2.16(d). If a minterm is covered by a prime implicant, a cross is made in the appropriate entry. Note that prime implicants are arranged in ascending number of literals.

It is now possible, by using the concepts of essential rows, row dominance, and column dominance, to select a minimal expression for the switching function [3]. Briefly, in our present context, a row is said to be essential if one or more of the minterms is covered only by the prime implicant in that row, so that this prime implicant has to be included in the final expression. A row Ri is said to be dominated by another row Rj if they correspond to p.i's with the same number of literals and the set of minterms covered by row Ri is a subset of the minterms covered by the other row Rj. In this case using Ri instead of Rj can be warranted as beneficial. A column Ci is said to be dominated by another column Cj if the set of prime implicants covering column Cj contains the set of prime implicants covering the other column Ci. In this case column Cj will be automatically covered if Ci is. For example, the two prime implicants selected are x_3x_4' and $x_1'x_2x_4$ and hence the minimal expression for the switching function

$$T(x_1,x_2,x_3,x_4) = \Sigma(2,5,6,7,10,14)$$

is again $x_3x_4' + x_1'x_2x_4$. For a more detailed description of the above procedure, see references [3,4,5,7].

Just as in the case with the mapping method, the above procedure can also be extended to include don't-care conditions as well.

The procedures described in this section are for designing AND-OR circuits. A procedure for designing NAND or NOR circuits will be given in the next section.

2.6 DESIGN OF NAND (NOR) CIRCUITS

It has been mentioned in Section 2.3 that circuits in which transistors are used exhibit the characteristics of NAND and NOR logic gates. Since transistors are now used in almost all logic circuits, it is necessary to look into the design of NAND (NOR) circuits.

First of all, let us consider the two-level AND-OR circuit shown in Fig. 2.17(a). The output from the OR-gate is given by

$$\begin{aligned} T(x_1,x_2,x_3) &= x_1x_2 + x_1'x_3' + x_2x_3 \\ &= ((x_1x_2)'(x_1'x_3')'(x_2x_3)')' \end{aligned}$$

from de Morgan's Theorem. It can easily be seen that the two-level NAND circuit shown in Fig. 2.17(b) realizes the same function. The inputs to both circuits are the same and the circuit in Fig. 2.17(b) can be obtained directly from that in Fig. 2.17(a) by simply replacing each of the AND-gates and OR-gates by a NAND-gate. This is true for all 2-level AND-OR and NAND circuits and this fact can be used conveniently in the design of 2-level NAND circuits: a 2-level AND-OR circuit is first designed using the procedure discussed in Section 2.5. The resulting circuit is then converted to a 2-level NAND circuit by simply replacing each of the AND-gates and OR-gates by a NAND gate.

In a similar fashion, a two-level OR-AND circuit can be converted into an equivalent NOR circuit by direct replacement of gates.

Unfortunately, the above discussion is only limited to two-level circuits, and cannot be extended directly to circuits of more than two levels. However, a lot of work has been done in the area dealing with the design of NAND (NOR) circuits and many systematic design procedures have been developed. We illustrate one of them through the use of an example. New definitions are introduced as they are needed. For a more detailed description of the procedure, see Torng [2,9].

Consider the implementation, using NAND gates, of the following switching function:

$$T(x_1,x_2,x_3,x_4) = \Sigma(0,1,2,3,5,6,11,12,13)$$

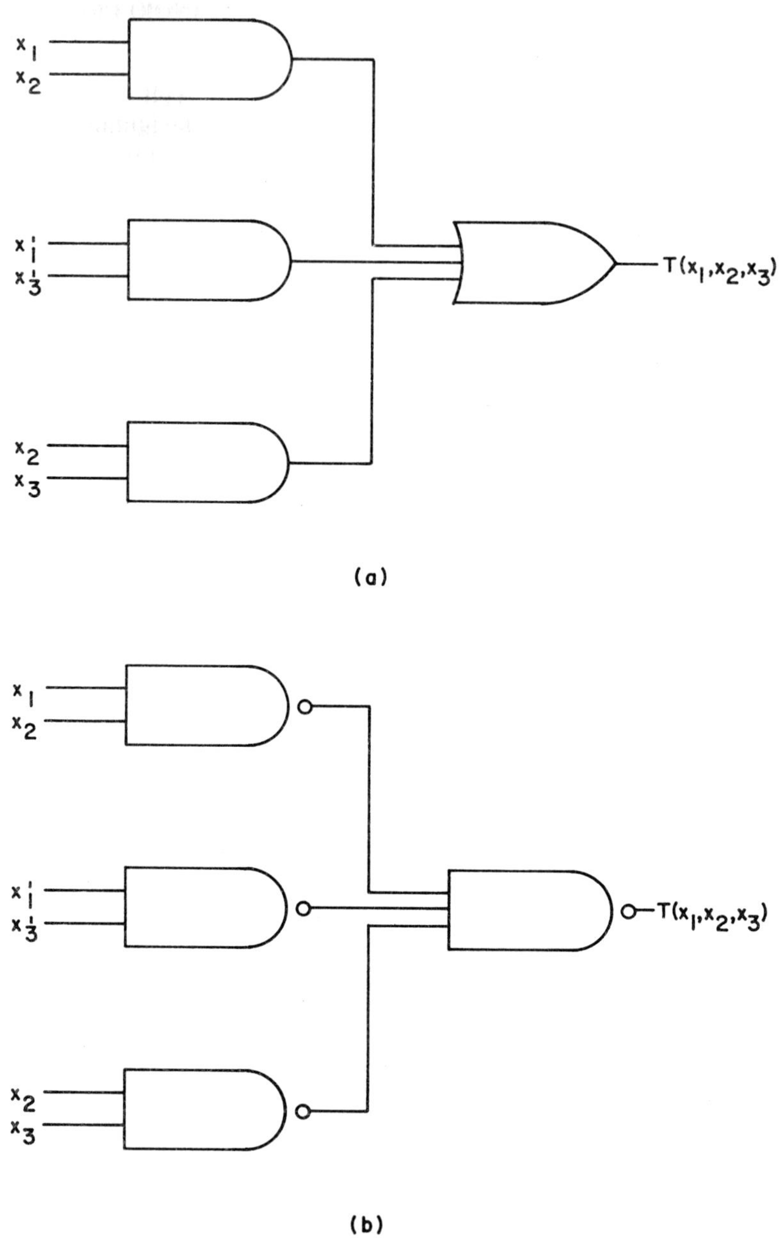

Figure 2.17 Relationship between two-level AND-OR circuits and NAND circuits

Step 1: First the function is plotted on a Karnaugh map in Fig. 2.18(a).

Step 2: The primary cube (the 0-cube corresponding to the combination of all 1s for the switching variables) is identified. Thus in the example, the primary cube is that corresponding to the variable vector 1111 and has an entry of 0.

Step 3: Gates with appropriate inputs to cover the cubes that can be combined with the primary cube are listed. In the example, the 0-cubes that can be combined with the primary cube 1111 are 0111 and 1110, as shown in Fig. 2.18(b). The resulting gates with the appropriate inputs are listed in Fig. 2.19(a). Note that the decimal codes corresponding to the input vectors for which the output is 0 are listed (I) and there is an entry *S* which indicates the value of the entries of corresponding 0-cubes in the Karnaugh map. Note also that the level of the gates (in this step, level 1) is indicated in the naming of the gates.

Step 4: New cubes are combined with previously combined cubes and gates are obtained as described in Step 3. This step is repeated alternately with cubes having 1-entries and then with cubes having 0-entries until the entire collection of cubes has been covered. The combining process of this step for the example is illustrated in the maps shown in Fig. 2.18(c)–(e), and the gates are listed in Fig. 2.19(b)–(d).

Step 5: Gates of levels 1 and 2 are now combined as shown in Fig. 2.20(a). Note that a gate in level 1 is connected to the input of a gate in level 2 only if there are common terms in the *I* sets of the two gates. The *I* sets of the gates of level 2 are now adjusted cancelling those terms that are contained in the *I* set of the input gates. Thus the *I* set of gate (2,1) is now changed from {12, 13,14,15} to {12,13} because gate (1,2) has an *I* set of {14,15}.

Step 6: Gates of levels 3 and 4 are now added in a similar fashion. The final circuit is shown in Fig. 2.20(b). Note that an additional gate (5,1) is needed to effect the required inversion to obtain the correct output.

In the above design procedure illustrated through the example, the primary cube chosen is 1111, as a result of which all the primary input lines are in an uncomplemented form. By choosing a different 0-cube as the primary cube, it is possible to design a circuit in which some of the primary inputs are in complemented forms. Thus, if in the same example, the input

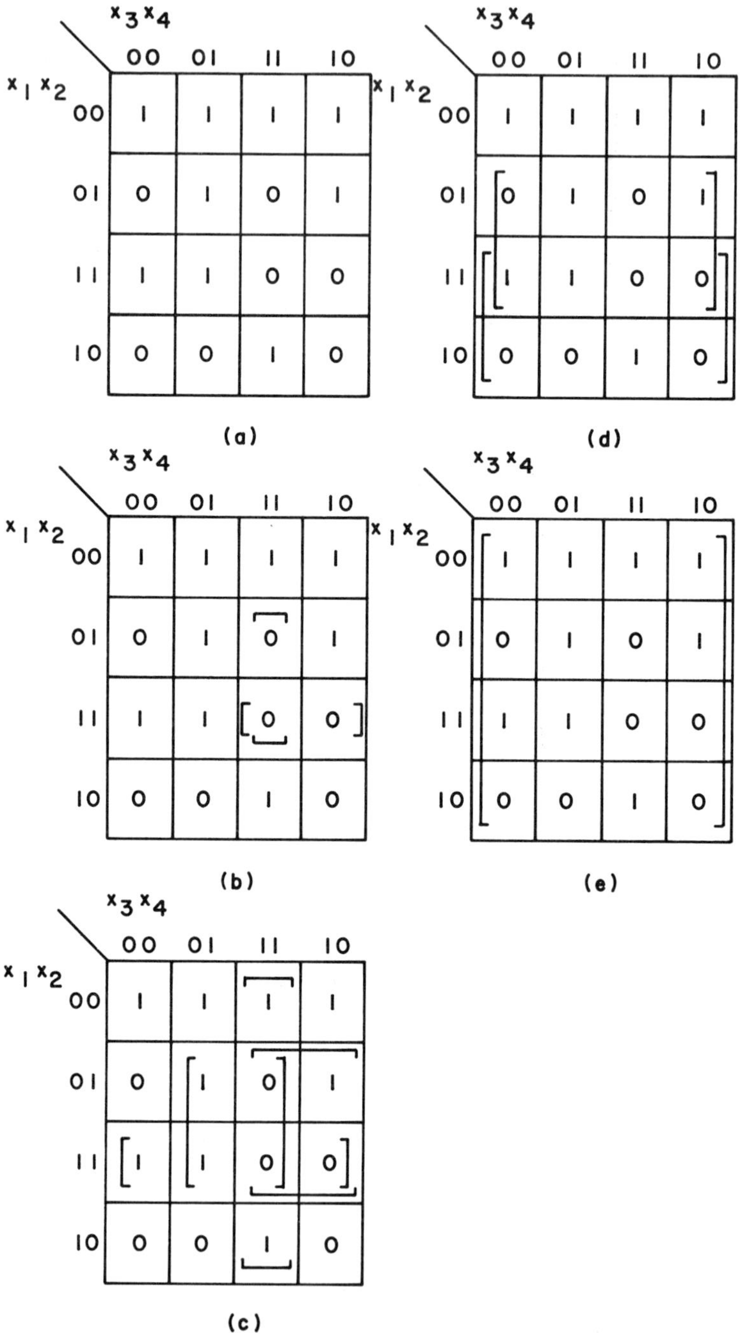

Figure 2.18 Example

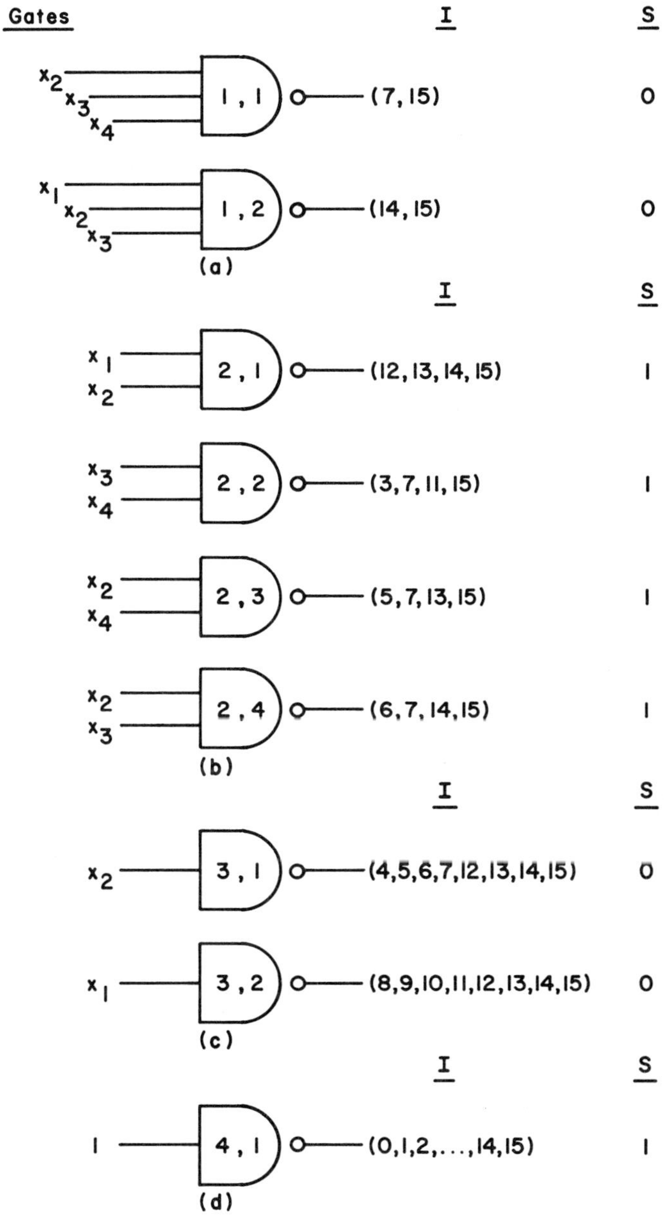

Figure 2.19 List of gates

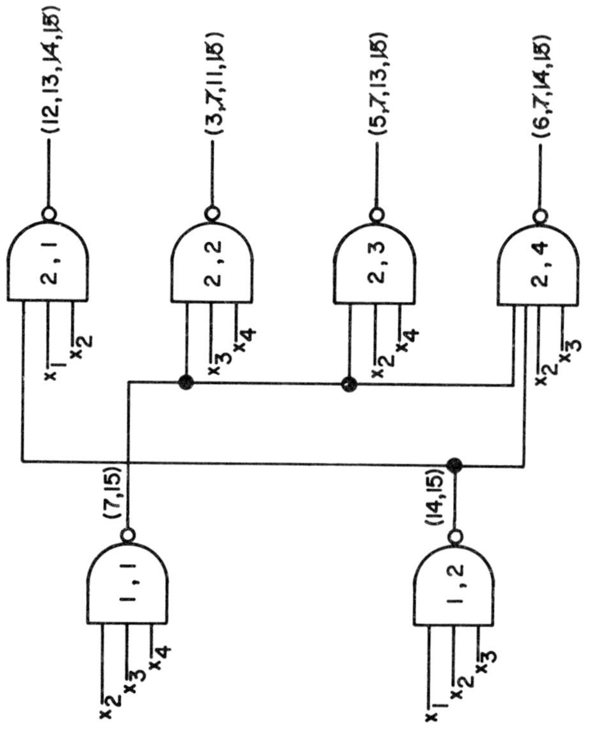
2,1
2,2
2,3
2,4
1,1
1,2
(12,13,14,15)
(3,7,11,15)
(5,7,13,15)
(6,7,14,15)
(7,15)
(14,15)
x1
x2
x3
x4

(a)

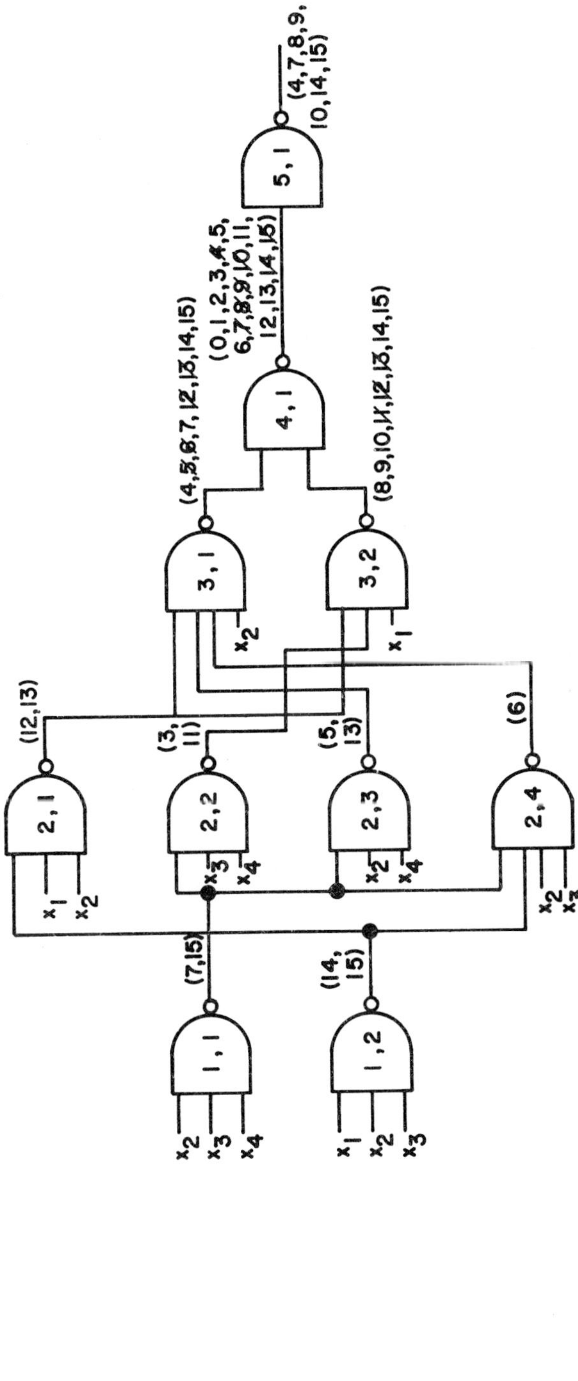

(b)

Figure 2.20 Combining gates

for the variable x_2 is in complemented form, so that only x_2' is available as a primary input, then we have to choose the 0-cube 1011 as the primary cube.

Circuits obtained using the above procedure are usually of more than two levels. Additional techniques can be developed to reduce the number of levels to at most three. In addition, it is also possible to reduce the number of NAND gates used in a circuit by eliminating redundant ones. The reader is referred to Torng [9] for a detailed description of the techniques required for both level-reduction and gate-reduction processes.

Because of the duality property, the above procedure can readily be modified to be used to design NOR circuits by simply interchanging + and ., 1 and 0 and replacing NAND gates by NOR gates.

The reader is referred to references given at the end of this chapter for numerous other procedures for the design of NAND (NOR) circuits, some of them using cells (a collection of gates) instead of gates as building units, a concept which is essential when using MOS circuits.

So far we have only considered the realization of individual functions. We shall now consider the realization of sets of functions.

2.7 MULTIPLE-OUTPUT CIRCUIT DESIGN

Consider a multiple-output circuit shown in block diagram form in Fig. 2.21. In the circuit there is more than one output line. Thus the circuit actually realizes as many subfunctions as the number of output lines. The subfunctions can, of course, be realized independently and the resulting circuits can then be combined together. However, if this is done, then even if the individual circuits are minimal, the combined circuit will often not be. It is therefore necessary to consider the realization of all the output subfunctions in one design procedure.

In general, what the technique tries to accomplish is to generate product terms which can be shared among several subfunctions.

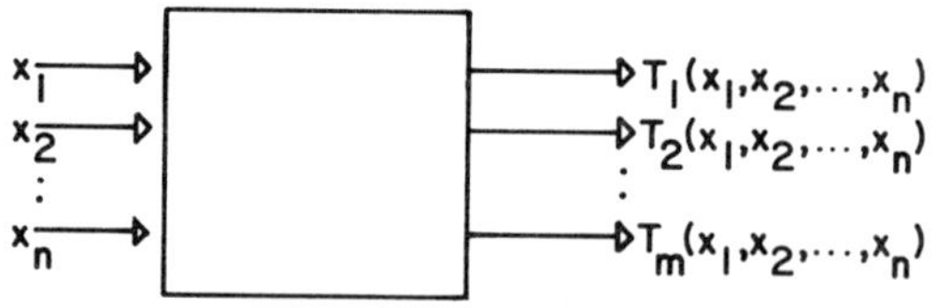

Figure 2.21 Multiple-output circuit

Let us consider the realization of the following multiple-output function:

$$T_1(x_1,x_2,x_3,x_4) = \Sigma(0,1,2,4,5,6,10,14)$$

$$T_2(x_1,x_2,x_3,x_4) = \Sigma(0,4,10,14)$$

$$T_3(x_1,x_2,x_3,x_4) = \Sigma(1,2,5,6)$$

The tabulation method as presented in Section 2.5 produces the groups of minterms shown in Fig. 2.22(a), where the rightmost entry S, in parentheses, identifies the subfunction or subfunctions in which the minterm is included. A new set of groups is now established as shown in Fig. 2.22(b). Note that the subfunction number is entered in S for a term resulting from the combination step only if both the combined terms are included in the subfunction. Thus only a 1 is entered for the term 00-0 even though the 0000 is a minterm for subfunctions T_1 and T_3. Note further that a term in Fig. 2.22(a) is checkmarked only if its S entry is identical to that of the combined term although, in the example, it so happens that every one of the minterms are checkmarked. Further combination of the new groups results in those shown in Fig. 2.22(c). Since no further combination is now possible, the combining process is terminated.

The prime implicant covering chart is shown in Fig. 2.22(d) and use of the techniques of essential rows, column dominance, and row dominance [3] allows us to obtain the following solution:

$$T_1(x_1,x_2,x_3,x_4) = x_1'x_3'x_4' + x_3'x_4 + x_1'x_3x_4' + x_1x_3x_4'$$

$$T_2(x_1,x_2,x_3,x_4) = x_1'x_3'x_4' + x_1x_3x_4'$$

$$T_3(x_1,x_2,x_3,x_4) = x_1'x_3'x_4 + x_1'x_3x_4'$$

Note that if the three subfunctions had been implemented independently, then T_1 would have been given the following simpler form

$$x_3x_4' + x_1'x_3'$$

but the overall circuit would require more gates for realization.

The circuit for the realization of the subfunctions is shown in Fig. 2.23.

	x_1	x_2	x_3	x_4	S	
0:	0	0	0	0	(1,2)	✓
1:	0	0	0	1	(1,3)	✓
2:	0	0	1	0	(1,3)	✓
4:	0	1	0	0	(1,2)	✓
5:	0	1	0	1	(1,3)	✓
6:	0	1	1	0	(1,3)	✓
10:	1	0	1	0	(1,2)	✓
14:	1	1	1	0	(1,2)	✓

(a)

	x_1	x_2	x_3	x_4	S	
0,1:	0	0	0	—	(1)	✓
0,2:	0	0	—	0	(1)	✓
0,4:	0	—	0	0	(1,2)	
1,5:	0	—	0	1	(1,3)	
2,6:	0	—	1	0	(1,3)	
2,10:	—	0	1	0	(1)	✓
4,5:	0	1	0	—	(1)	✓
4,6:	0	1	—	0	(1)	✓
6,14:	—	1	1	0	(1)	✓
10,14:	1	—	1	0	(1,2)	

(b)

	x_1	x_2	x_3	x_4	S
0,1,4,5:	0	—	0	—	(1)
0,2,4,6:	0	—	—	0	(1)
2,6,10,14:	0	—	—	0	(1)

(c)

	T_1								T_2				T_3			
	0	1	2	4	5	6	10	14	0	4	10	14	1	2	5	6
x_3x_4'			x			x	x	x								
$x_1'x_4'$	x		x	x		x										
$x_1'x_3'$	x	x		x	x											
$x_1'x_3'x_4'$	x			x					x	x						
$x_1'x_3'x_4$		x			x								x		x	
$x_1'x_3x_4'$			x			x								x		x
$x_1x_3x_4'$							x	x			x	x				

(d)

Figure 2.22 Tabulation method for multiple-output design

2.8 FUNCTIONAL DECOMPOSITION

Theoretically, a gate can accept any number of inputs. In practice, however, the maximum number of inputs that can be fed into a gate is limited by its physical characteristics (this number is commonly known as its fan-in) and it may then become necessary to use circuits with more than two levels. One approach to deriving realizations with more than two levels is functional decomposition.

The functional decomposition of a switching function $T(x_1,x_2,\ldots,x_n)$ transforms it into the form

$$T(x_1,x_2,\ldots,x_n) = F(\phi(y_1,y_2,\ldots,y_s),z_1,z_2,\ldots,z_r)$$

where $1 < s \leq n - 1$, and $\{y_1.y_2,\ldots,y_s\} \cup \{z_1,z_2,\ldots,z_r\} = \{x_1,x_2,\ldots,x_n\}$. If the sets $\{y_1,y_2,\ldots,y_s\}$ and $\{z_1,z_2,\ldots,z_r\}$ are in addition disjoint, i.e., we have $r = n - s$, then the decomposition is said to be *disjunctive*. For simplicity we limit our discussion to disjunctive decomposition only. A function which can be decomposed in such a manner is said to be functionally decomposable. One of the reasons for introducing the concept of functional decomposition is that an expression obtained through decomposition can usually be realized by a circuit which requires fewer input lines, and this special feature can quite often prove to be valuable. Another reason is that functional decomposition sometimes enables the design of a complex circuit to be decomposed into the design of a number of simpler and/or standard subcircuits (see the next section and Chapter

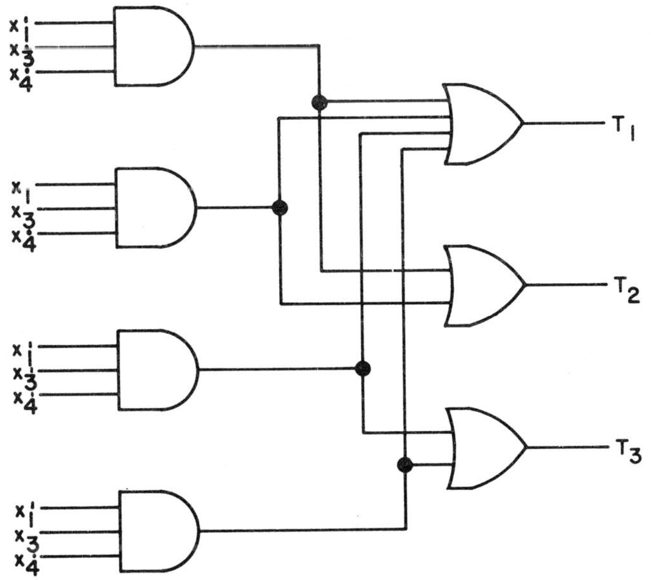

Figure 2.23 Multiple output circuit

3). The block diagram in Fig. 2.24 illustrates the process of functional decomposition.

Example: The function

$$T(x_1,x_2,x_3,x_4) = x_1x_2 + x_2x_3 + x_1'x_3'x_4$$

can be functionally decomposed into

$$T(x_1,x_2,x_3,x_4) = \phi x_2 + \phi' x_4$$

where $\phi(x_1,x_3) = x_1 + x_3$, as shown in Fig. 2.25.

Example: The function which describes a full adder

$$T(x_1,x_2,x_3) = x_1'x_2'x_3 + x_1'x_2x_3' + x_1x_2'x_3' + x_1x_2x_3$$

can be decomposed into

$$T(x_1,x_2,x_3) = \phi x_1' + \phi' x_1$$

where $\phi(x_2,x_3) = x_2'x_3 + x_2x_3'$, as shown in Fig. 2.26.

A procedure for decomposing a switching function will now be established. The decomposition of a switching function $T(x_1,x_2,\ldots,x_n)$ into $F(\phi(y_1,y_2,\ldots,y_s), z_1,z_2,\ldots,z_{n-s})$ induces a two-block partition P on the set of switching variables $(x_1,x_2,\ldots,x_n)$ where

$$P = \{\overline{y_1,y_2,\ldots,y_s};\ \overline{z_1,z_2,\ldots,z_{n-s}}\}$$

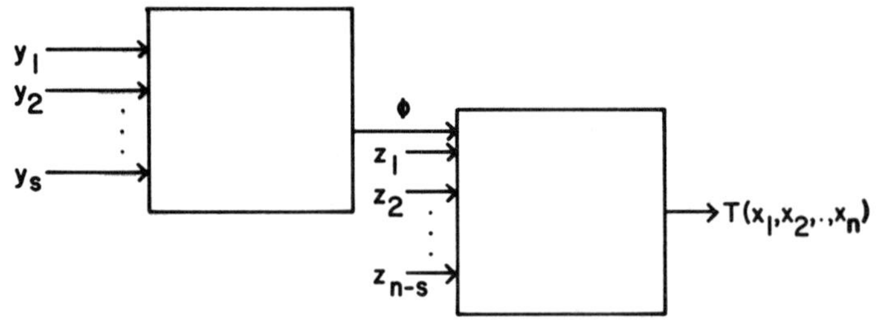

Figure 2.24 Functional decomposition

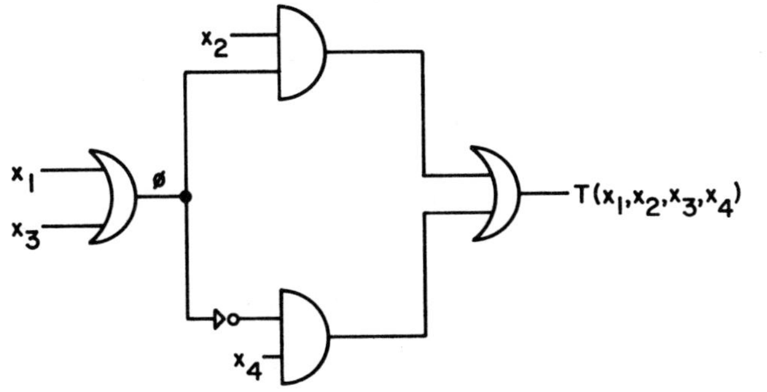

Figure 2.25 $T(x_1,x_2,x_3,x_4) = x_1x_2 + x_2x_3 + x_1'x_3'x_4.$

The partition P can be used to obtain a $2^{n-s} \times 2^s$ matrix representation of the switching function T. The rows enumerate the 2^{n-s} possible combinations of values of the variables $z_1,z_2,\ldots,z_{n-s}$ arranged in their decimal order and the columns enumerate the 2^s possible combinations of values of the variables $y_1,y_2,\ldots,y_s$ also arranged in their decimal order. An entry of a 1 in the matrix is made if the corresponding minterm defined by the y and z is covered by the switching function and an entry of a 0 in the matrix is made otherwise.

Example: A partition matrix for the function $T(x_1,x_2,x_3,x_4) = x_1x_2 + x_2x_3 + x_1'x_3'x_4$ for the partition $\{\overline{x_1,x_3}; \overline{x_2,x_4}\}$ is shown in Fig. 2.27.

The row multiplicity μ of a matrix formed in the above fashion is defined to be the number of distinct rows and the column multiplicity ν of the

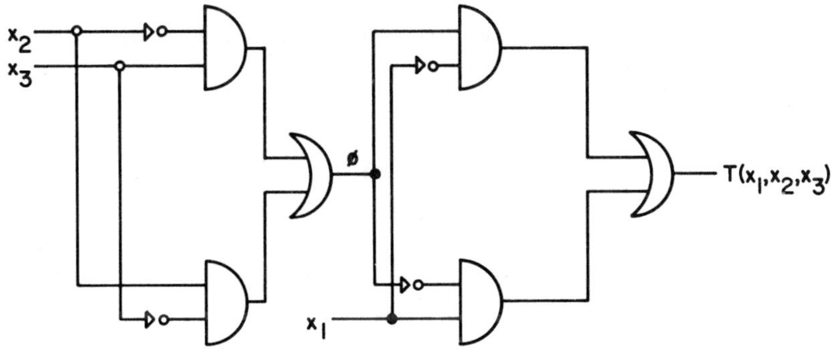

Figure 2.26 $T(x_1,x_2,x_3) = x_1'x_2'x_3 + x_1'x_2x_3' + x_1x_2'x_3' + x_1x_2x_3$

x_2x_4 \ x_1x_3	00	01	10	11
00	0	0	0	0
01	1	0	0	0
10	0	1	1	1
11	1	1	1	1

Figure 2.27 Partition matrix

matrix is similarly defined to be the number of distinct columns in the matrix. In the matrix in Fig. 2.27, $\mu = 4$ and $\nu = 2$.

The entire decomposition procedure is based on the following well-known theorem:

Theorem 2.2 A switching function $T(x_1,x_2,\ldots,x_n)$ is decomposable such that $T(x_1,x_2,\ldots,x_n) = F(\phi(y_1,y_2,\ldots,y_s), z_1,z_2,\ldots,z_{n-s})$ if and only if $\nu \leq 2$ for the matrix corresponding to the partition

$$\{\overline{y_1,y_2,\ldots,y_s};\ \overline{z_1,z_2,\ldots,z_{n-s}}\}$$

For a proof of the above theorem, see References 5 and 9.

Thus $\nu \leq 2$ is a necessary and sufficient condition under which a function is decomposable according to the particular partition on the switching variables. Since $\nu = 2$ for the matrix in the above example, the particular function

$$T(x_1,x_2,x_3,x_4) = x_1x_2 + x_2x_3 + x_1'x_3'x_4$$

is indeed decomposable as has been shown previously.

Since the roles of rows and columns in a partition matrix can be interchanged to effect only a change in the ordering of the two blocks in the partition matrix, it should be evident that a switching function is also decomposable when $\mu \leq 2$. In this case, it will be possible to obtain the following decomposition:

$$T(x_1,x_2,\ldots,x_n) = F_1(\phi(z_1,z_2,\ldots,z_{n-s}),y_1,y_2,\ldots,y_s)$$

Thus the procedure for obtaining a decomposition for a switching function $T(x_1,x_2,\ldots,x_n)$ corresponding to the partition $\{\overline{y_1,y_2,\ldots,y_s};\ \overline{z_1,z_2,\ldots,z_{n-s}}\}$ consists of the following steps:

Step 1: A partition matrix is obtained as described in the above.

Step 2: Test to see if $v \leq 2$. If yes, go to step 3; if no, we conclude that T cannot be decomposed into $F(\phi(y_1,y_2,\ldots,y_s), z_1,z_2,\ldots,z_{n-s})$ and exit.

Step 3: A nontrivial row, that is, a row which is a mixture of 0s and 1s, is chosen to be ϕ. Next the remaining nontrivial row, if any, is designated to be ϕ'. Then a row of all 0s is designated to be 0 and a row of all 1's is designated to be 1. Thus ϕ has already been defined in this step.

Step 4: The minterms in the function $F(\phi,z_1,z_2,\ldots,z_{n-s})$ are now determined. Let m_i be a minterm with a decimal representation of i. Then depending on whether row i is ϕ, ϕ', 0 or 1, m_i is a minterm or not a minterm in F according to Table 2.5, where a 1 stands for inclusion and a 0 stands for exclusion. Note that $m_{i+2^{n-s}}$ is a minterm with ϕ as the first literal and m_i is that with ϕ' as the first literal. F is then obtained by summing up all the included minterms.

Example: Let us use the above procedure to find a decomposition for the function given by:

$$T(x_1,x_2,x_3,x_4) = x_1x_2 + x_2x_3 + x_1'x_3'x_4$$

for the partition $\{\overline{x_1,x_3}; \overline{x_2,x_4}\}$.

Step 1: The partition matrix has been found in a previous example (Fig. 2.27).

Step 2: Since $v = 2$ a decomposition exists.

Step 3: Let row 2 be ϕ. Thus we have $\phi = x_1 + x_3$. Row 1 is then ϕ', row 0 is 0, row 3 is 1.

Row	m_i	$m_{i+2^{n-s}}$
0	0	0
1	1	1
ϕ	0	1
ϕ'	1	0

Table 2.5 Table to find $F(\phi,z_1,z_2,\ldots,z_{n-s})$.

Step 4: The set of minterms included is $\{m_1,m_6,m_3,m_7\}$. Nore that $2^{n-s} = 4$. Hence

$$F(\phi,x_2,x_4) = m_1 + m_6 + m_3 + m_7$$
$$= \phi'x_2'x_4 + \phi x_2x_4' + \phi'x_2x_4 + \phi x_2x_4$$
$$= \phi x_2 + \phi' x_4$$

A repeated application of the above procedure for each distinct partition can of course be used to test whether a certain switching function is decomposable at all. In such a case, the fact that a partition matrix can also be looked at sideways should be taken into account to reduce the number of partition matrices needed to be formed.

2.9 OTHER METHODS OF IMPLEMENTATION

One application of functional decomposition is in the realization of functions using commercially available integrated circuits (ICs). Many commonly used functions have been implemented as single-chip integrated circuits. These chips have a wide range of complexity in the number of gates involved. The simplest can be made up of a single logic gate and the more complex one can realize an entire central processing unit (CPU). More will be said about ICs in a later chapter.

In this section, we present a number of other methods that can be used to implement a switching function. The first method uses a data selector or *multiplexer* instead of discrete logic gates to implement a given switching function. Multiplexers (see Problem 2.21 for the design of a multiplexer) are widely used in the design of digital systems and are available as low-cost IC chips and hence the implementation method given here is of tremendous practical use.

A block diagram of an n-input, m address multiplexer is shown in Fig. 2.28. It has a number of input lines $x_0,x_1,\ldots,x_{n-1}$ and a number of select lines $z_1,z_2,\ldots,z_m$ where $n = 2^m$. There is also a single output line x. For convenience, let us use the particular case with $n = 4$ in the following discussion. The reader should have no difficulty in making the generalization.

Signals on the select lines control the connection of the input lines to the output line. If both select inputs are zeros, the output is x_0. Similarly, for $z_1z_2 = 01$, we have $x = x_1$, for $z_1z_2 = 10$ we have $x = x_2$ and for $z_1z_2 = 11$, we have $x = x_3$. Thus the operation can be represented by the following switching function:

$$x = x_0z_1'z_2' + x_1z_1'z_2 + x_2z_1z_2' + x_3z_1z_2$$

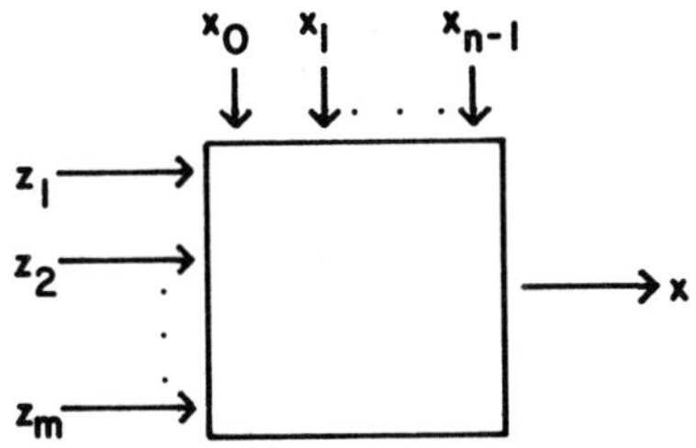

Figure 2.28 Block diagram of multiplexer

To generate a given function of three variables using the multiplexer, it is necessary to have the complement of only one of the three variables and the constants 1 and 0 to act as inputs to the lines x_0,x_1,x_2 and x_3. The other two variables, in uncomplemented forms, are connected to z_1 and z_2. For determination of the inputs to lines x_0,x_1,x_2 and x_3, the function is first entered on a Karnaugh map and then each input line is associated with two 0-cubes of the map. The association depends on the choice of the two variables connected to lines z_1 and z_2 but is independent of the particular function being entered on the map. This is best illustrated by means of an example.

Suppose we want to generate the following function using a multiplexer:

$$f(a,b,c) = \Sigma(2,3,4,6)$$

The function is entered in a Karnaugh map as shown in Fig. 2.29(a) and if we connect the two variables c and b to z_1 and z_2 respectively, then the corresponding association of the pairs of 0-cubes is that shown in Fig. 2.29(b).

It is then easy to obtain from Fig. 2.29(b) the appropriate inputs to the four lines x_0,x_1,x_2 and x_3. To obtain the input for line x_0, for example, we note that the output should be exactly identical to the input a, i.e. $a = 0$ produces an output of 0 and $a = 1$ produces an output of 1. Hence, the input at x_0 should be a. Using similar arguments, we obtain the circuit in Fig. 2.30 as an implementation of the given function. The reader can verify the correctness of the circuit.

It can be seen that a 4-input multiplexer can be used to generate any 3-variable function. Similarly an 8-input multiplexer can be used to generate any 4-variable function. Although we can theoretically go on with this generalization indefinitely, multiplexers of greater than eight input lines are usually built from smaller ones and therefore their use in the

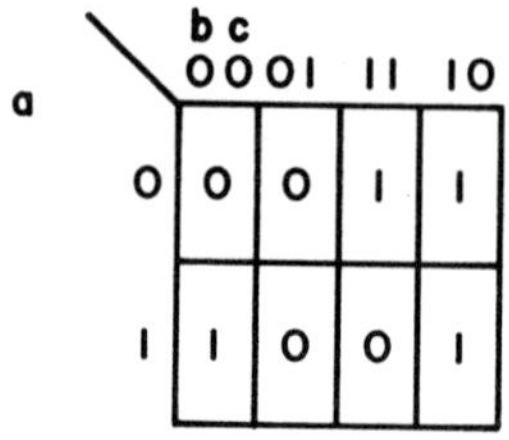

a.

(a) Karnaugh map of $f(a,b,c)$

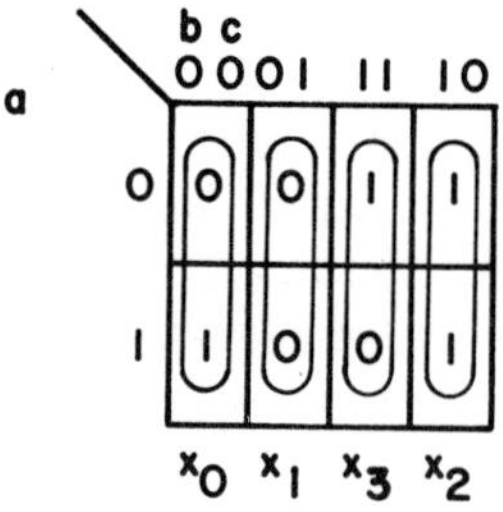

b.

(b) Karnaugh map of associated cubes

Figure 2.29

implementation of functions of more than four variables is not as straightforward. For more discussions on the use and design of multiplexers, see Reference 1.

Another increasingly popular method that can be employed to implement a switching function is the use of a read-only-memory (ROM). A

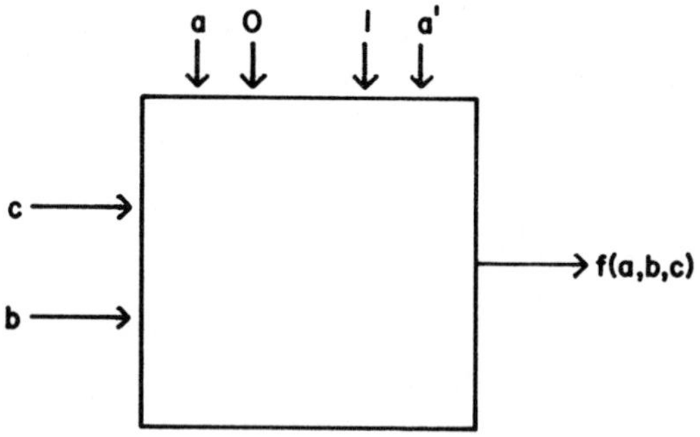

Figure 2.30 Implementation using multiplexer

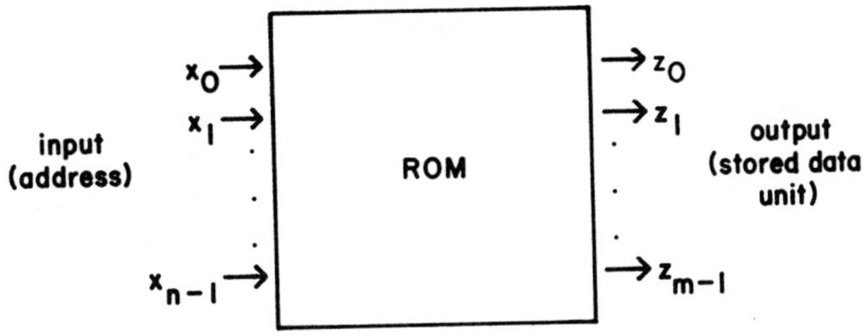

Figure 2.31 Schematic diagram of a ROM

ROM is a memory circuit that can store a certain amount of information and will be discussed in more detail in Chapter 4. For the moment, it suffices to look at a ROM as a black box containing some previously fed in data permanently stored in it. Each unit of data can be read out by the application of a particular set of input address signals (Fig. 2.31). Thus if there are n input lines, then a maximum of 2^n units of data can be stored and accessed.

The switching function

$$f(a,b,c) = \Sigma(2,3,4,6)$$

can be implemented using a ROM having three input lines and capable of storing eight bits of data. The data are stored in the eight locations as shown in Table 2.6. As a result for example, application of the input combination 100 will fetch the contents 1 and make it available at the output terminal. It is easy to see that we have stored the function's entire truth table as data in the ROM! Because of this ease of application in

address *abc*	content
000	0
001	0
010	1
011	1
100	1
101	0
110	1
111	0

Table 2.6 Data stored in ROM

small problems, ROMs are becoming more and more popular as their costs keep decreasing. We shall see in Chapter 7 how ROMs can be used to design CPUs although not in such a straightforward fashion.

Switching functions can also be implemented using programmable logic arrays (PLAs). PLAs are very similar to ROMs except that while the ROMs work with minterms, the PLAs work with product terms which may not be minterms. Thus for the same memory size, a PLA can implement a switching function of a larger number of variables [1].

2.10 REGULAR CIRCUITS

A number of the most commonly used circuits in the design of a computer have very regular structures; certain subcircuits in each circuit are either exactly duplicated a number of times or have very similar forms to those of some other subcircuits. For example, an adder that is used to perform addition on 32-bit numbers can be implemented as an array of 32 full adder subcircuits (for the design of a full adder, see Problem 2.7). Some of the outputs of each subcircuit may be used as inputs to other subcircuits but the role of each subcircuit will be very regular. Because of the usually large number of total inputs and outputs, the conventional methods of implementation of such circuits as single circuits becomes very difficult if not impossible. In this section, we discuss briefly through the use of examples how these regular circuits can be more conveniently designed. Exact methods differ for different circuits but the general approach is the same—use decomposition and proceed by inspection.

When a computer has completed an operation, the numerical result is usually given in binary form. Since most people prefer to work in decimal form it is necessary to decode the output. One method is to use a decoder such as the one shown in Fig. 2.32.

The circuit in Fig. 2.32 is obtained by observing that there is only one input combination for which a particular output terminal will assume a value of 1. So each output function can be generated through the use of only one single AND gate.

Another example is the design of an encoder for generating 3-bit binary numbers. By observing that only one of the 8 input signals is a 1 at any one time and that each of the 3 outputs assumes a value of 1 for only four of the eight possible inputs, the regular circuit shown in Fig. 2.33 can easily be obtained.

Most regular circuits can be decomposed and implemented as arrays of simpler subcircuits which can then be designed using the methods we

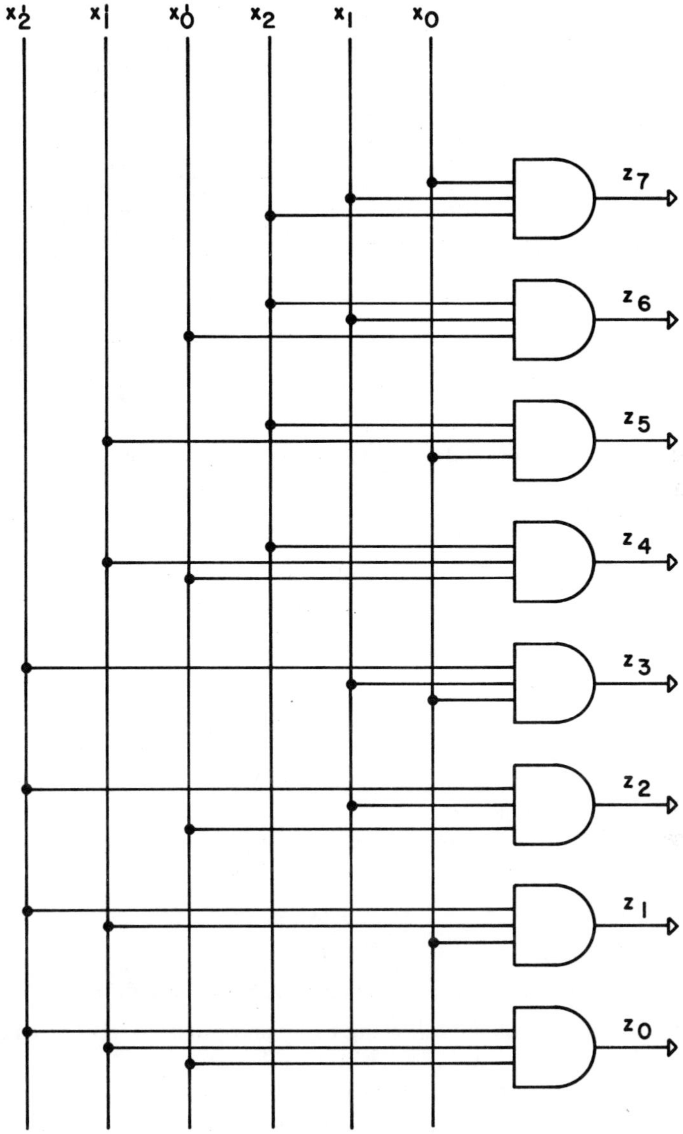

Figure 2.32 A decoder for 3-bit binary numbers

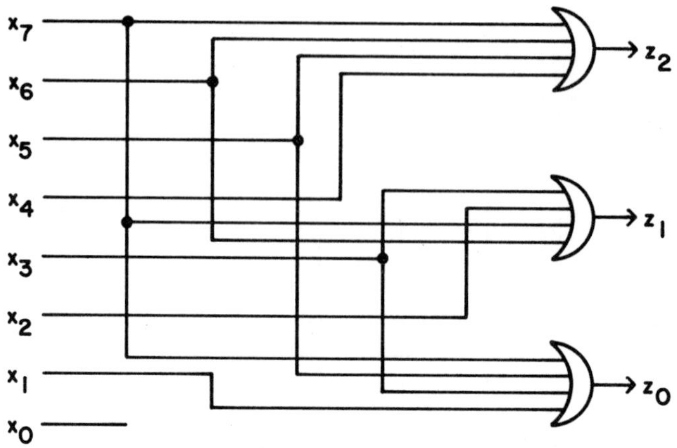

Figure 2.33 An encoder for 3-bit codes

have described in the previous sections. A number of these regular circuits will be discussed in more detail in later chapters.

PROBLEMS

2.1 Find the equivalent expressions with fewest literals for each of the following:
(a) $xyz + xyz' + xy'z + xy'z' + x'z$
(b) $xy' + x'y'z + y'z'$
(c) $x_1 + x_1'x_2 + x_1'x_2'x_3 + x_1'x_2'x_3'x_4 + \cdots$

2.2 Show that
$(x + y)(x' + z)(y + z) = (x + y)(x' + z)$

2.3 Prove Shannon's Expansion Theorem.

2.4 Let f_d be the dual function of function f. Show that $f_d = f'(x_1', x_2', \ldots, x_n')$.

2.5 Implement each of the following using an optimal AND-OR circuit:
(a) $T(x_1, x_2, x_3) = x_1'x_2'x_3' + x_1x_2'x_3' + x_1x_2'x_3$
(b) $T(x_1, x_2, x_3) = x_1x_2' + x_1'x_2'x_3 + x_2'x_3'$

2.6 Obtain optimal AND-OR circuits for the following functions:
(a) $T(x_1, x_2, x_3, x_4) = \Sigma(0,1,2,5,12,14,15)$
(b) $T(x_1, x_2, x_3, x_4) = \Sigma(1,2,3,5,7,9,15)$

2.7 Design a full adder with inputs x_1, x_2 and carry y and outputs sum s and carry c and realize it with a minimum number of AND, OR, and NOT gates.

2.8 Design a full-subtractor with inputs minuend x_1, subtrahend x_2, and borrow y; and outputs difference d, and borrow b, and realize it with a minimum number of AND, OR, and NOT gates.

2.9 By using the Karnaugh map, realize each of the following with a minimum number of AND and OR gates:
(a) $T(x_1,x_2,x_3,x_4) = \Sigma(0,4,6,10,11,13)$
(b) $T(x_1,x_2,x_3,x_4) = \Sigma(3,4,5,7,11,12,14,15)$
(c) $T(x_1,x_2,x_3,x_4) = \Sigma(4,5,7,12,14,15) + D(3,8,10)$
(d) $T(x_1,x_2,x_3,x_4) = \Sigma(0,4,5,8,12) + D(1,3,9,15)$
where the set in D is the set of don't-cares conditions.

2.10 Construct a Karnaugh map for the following 5-variable function:
$T(x_1,x_2,x_3,x_4,x_5) = \Sigma(1,2,4,5,6,11,12,13,14,15)$

2.11 Use the mapping method to realize the following with a minimum number of AND and OR gates:
$T(x_1,x_2,x_3,x_4,x_5) = \Sigma(1,2,9,13,27,28,30,31)$

2.12 Repeat Problem 2.11 using the tabulation method instead.

2.13 Use the tabulation method to realize the following with a minimum number of AND and OR gates:
$T(x_1,x_2,x_3,x_4) = \Sigma(1,2,4,5,6,11,12,13,14,15)$

2.14 Realize the function in Problem 2.9(a) with only NAND gates.

2.15 Realize the function in Problem 2.9(b) with only NAND gates.

2.16 Realize the following multiple-output function using a minimum number of gates:
$T_1(x_1,x_2,x_3,x_4) = \Sigma(0,1,3,4,8,11)$
$T_2(x_1,x_2,x_3,x_4) = \Sigma(4,5,10,11,12)$
$T_3(x_1,x_2,x_3,x_4) = \Sigma(0,4,10,12,14)$

2.17 Realize the following multiple-output function using a minimum number of gates:
$T_1(x_1,x_2,x_3,x_4) = \Sigma(0,6,7,8) + D(1,2)$
$T_2(x_1,x_2,x_3,x_4) = \Sigma(1,2,7,15) + D(5,6,8)$
$T_3(x_1,x_2,x_3,x_4) = \Sigma(0,1,2,7,8,15) + D(4,10,12)$

2.18 Decompose
$T(x_1,x_2,x_3,x_4,x_5) = \Sigma(1,2,7,11,13,14,19,23,27,29,30)$
into the form
$T(x_1,x_2,x_3,x_4,x_5) = F(\phi(x_1,x_2,x_3),x_4,x_5)$
by determining the function F and ϕ.

2.19 Identify all possible decompositions for the following function:
$T(x_1,x_2,x_3,x_4) = \Sigma(0,2,5,7,11,14)$

2.20 The following is a block diagram of an n-line priority encoder:

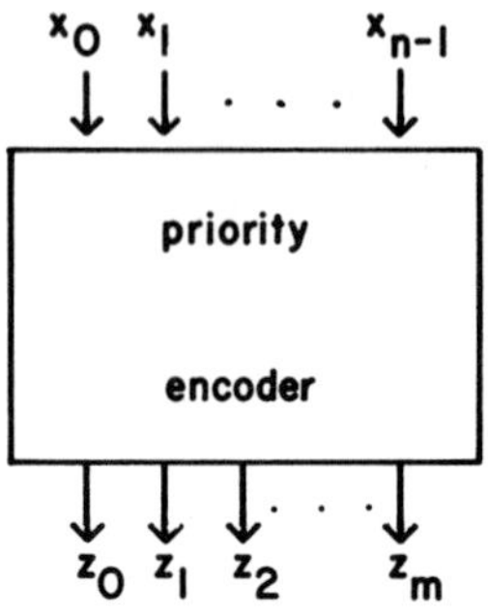

where $n = 2^m$. The function of the priority encoder is to detect the presence of a signal (usually representing requests) on the line x_i with the highest priority, where a line x_p has a higher priority over line x_q if $p < q$. The corresponding highest priority address i is displayed in binary coded form on output lines $z_1.z_2,...,z_m$. Output lines $z_0 = 0$ if there is no request on any of the lines and $z_0 = 1$ if there are one or more requests. Design a 4 line priority encoder. Design an 8-line priority encoder using two 4-line priority encoders and some auxiliary gates.

2.21 The following is a block diagram of a data selector (multiplexer):

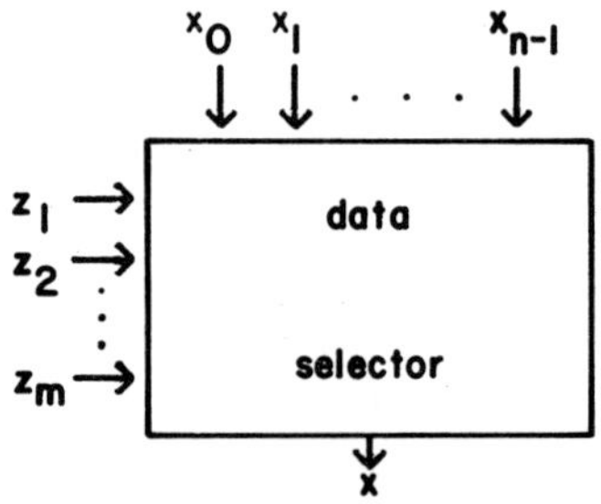

where $n = 2^m$. The data selector outputs x such that $x = x_i$ where i is the address indicated by the lines $z_1,z_2,...,z_m$. Design a data selector with $n = 8$.

2.22 Implement each of the following switching functions using a single multiplexer:

(a) $f(a,b,c) = \Sigma(1,2,4,7)$

(b) $f(a,b,c) = abc' + bc + a'b'c'$ and don't care $ab'c'$

REFERENCES

1. Blakeslee, T. R. *Digital Design with Standard MSI and LSI.* New York: John Wiley & Sons, 1975.
2. Gimpel, J. F. "The Minimization of TANT Networks." *IEEE Trans. on Elect. Comp.* vol. EC-16, Feb. 1967, pp. 18–38.
3. Gimpel, J. F. "A Reduction Technique for Prime Implicant Tables." *IEEE Trans. on Elect. Comp.* vol. EC-14, Aug. 1965, pp. 535–541.
4. Hill, F. J. and Peterson, G. R. *Introduction to Switching Theory and Logical Design.* New York: John Wiley & Sons, 1968.
5. Kohavi, S. *Switching and Finite Automata Theory.* New York: McGraw-Hill, 1965.
6. Liu, T. K. Synthesis Algorithms for 2-Level MOS Networks. *IEEE Trans. on Comp.* vol. C-24, Jan. 75, pp. 72–80.
7. McCluskey, E. J. *Introduction to the Theory of Switching Circuits.* New York: McGraw-Hill, 1965.
8. Schneider, P. R. and Dietmeyer, D. L. An Algorithm for Synthesis of Multiple-Outlet Combination Logic. *IEEE Trans. on Comp.* vol. C-17, Feb. 68, pp. 117–128.
9. Torng, H. C. *Switching Circuit Theory and Logic Design.* Reading MA: Addison-Wesley, 1972.

Chapter 3

SEQUENTIAL MACHINES AND DESIGN PROCEDURES

As indicated in the last chapter, a combinational circuit does not possess any memory, and hence, its output at any moment is dependent only on its input at the same moment. On the other hand, a sequential circuit constructed from combinational elements and storage elements has the capability of remembering some past history so its output may depend not only on the present input but also on a certain amount of stored information. The amount of information that can be stored depends on the number of available storage elements. In some sequential circuits, the memory capability is obtained from the inherent, small but finite delays in the combinational elements in the circuits and no distinguishable storage elements are otherwise present in the circuits. In this chapter, a finite-state machine is first defined and then the design of various types of sequential circuits is discussed and design procedures are established in the process. For more detailed discussions on the various subjects, the reader is referred to the references given at the end of this chapter.

3.1 FINITE-STATE MACHINE

A sequential machine is an abstract model which can be used to study sequential circuits. In this model, the possible inputs are assumed to be sequences of symbols selected from a finite set I of input symbols and the outputs are assumed to be sequences of symbols selected from a finite set Z of output symbols. The memory of the machine is represented by a set of states, one of which the machine can assume at any instance, and by a next-state function which defines the next state that the machine will assume as a function of its present state and input and an output function which indicates the output as a function of the present state and input. A sequential machine which has a finite number of states is known as a *finite-state machine*.

Formally, we define a finite-state machine M to be the 5-tuple

$M = (I,Z,Q,\delta,w)$ where
I is the set of input symbols,
Z is the set of output symbols,
Q is the finite set of states,
δ is the next-state function and is a mapping from $I \times Q$ into Q,
w is the output function and is a mapping from $I \times Q$ into Z.

A machine defined as above is known as a Mealy machine [2] to distinguish it from a machine in which the output function is a mapping from Q into Z rather than $I \times Q$ into Z which is known as a Moore machine [2]. If either of the mappings w or δ involves don't-care conditions, then the machine is known as an incompletely-specified machine. Section 3.4 will be devoted to the treatment of incompletely-specified machines and until then, w and δ will be assumed to involve no don't-care conditions. The Mealy and Moore machines are shown in block diagram forms in Fig. 3.1.

There are a number of other different ways of representing a finite-state machine M to facilitate ease of specification and analysis of the machine.

The second method displays the five items identifying the machine in a state table. Let $M1$ be given as:

$I = \{0,1\}$	$\delta(0,q_0) = q_0$	$\delta(1,q_0) = q_1$
$Z = \{0,1\}$	$\delta(0,q_1) = q_1$	$\delta(1,q_1) = q_2$
$Q = \{q_0,q_1,q_2,q_3\}$	$\delta(0,q_2) = q_2$	$\delta(1,q_2) = q_3$
	$\delta(0,q_3) = q_3$	$\delta(1,q_3) = q_3$
	$w(0,q_0) = 0$	$w(1,q_0) = 0$
	$w(0,q_1) = 0$	$w(1,q_1) = 0$
	$w(0,q_2) = 0$	$w(1,q_2) = 1$
	$w(0,q_3) = 0$	$w(1,q_3) = 0$

The state table for machine $M1$ is given in Table 3.1. In the state table, the present states, the next states and the outputs for various input combinations are tabulated in columns. Thus if there are n input variables, then there will be 2^n columns of entries for both the next states and the outputs (see Table 3.2).

Example: Table 3.2 shows the state table of a machine $M2$ having two input variables.

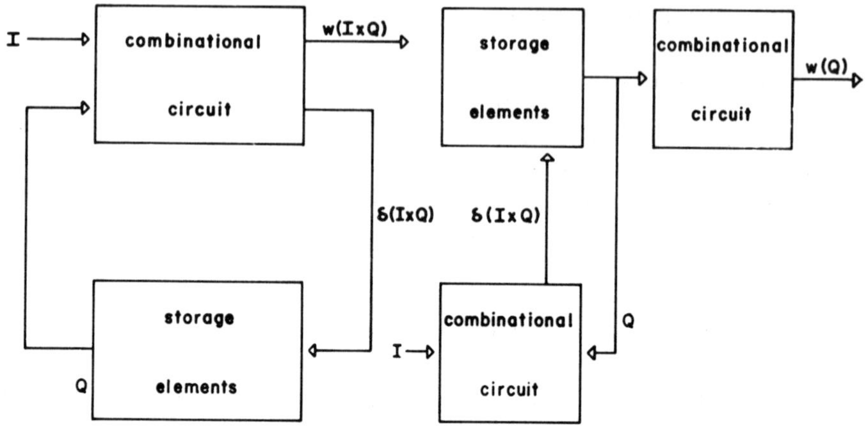

Figure 3.1 (a) Mealy machine (b) Moore machine

A state table expresses the behavior of a machine in a form which is very well suited for certain operations on the machine, such as machine reduction and machine equivalence testing, to be discussed in a later section. However, the dynamic behavior of a machine is much better exposed through the use of a state diagram, which provides a graphical representation of the operation of the machine. In a state diagram, each state is represented by a node and the state transitions are indicated by arcs. The following example illustrates the use of a state diagram to specify a finite-state machine.

Example: The state diagram for the machine $M1$ is shown in Fig. 3.2.

Example: The state diagram for machine $M2$ is shown in Fig. 3.3.

Present state	Next state value for input 0	1	Output for input 0	1
q_0	q_0	q_1	0	0
q_1	q_1	q_2	0	0
q_2	q_2	q_3	0	1
q_3	q_3	q_3	0	0

Table 3.1 State table for $M1$

Present state	Next state 00	01	11	10	Output 00	01	11	10
q_0	q_0	q_1	q_0	q_0	0	0	0	0
q_1	q_2	q_0	q_1	q_1	0	1	0	0
q_2	q_2	q_0	q_2	q_2	0	1	0	0

Table 3.2 State table for *M*2

Note that in a state diagram, an input combination together with the output is drawn next to the corresponding arc indicating the associated state transition.

A third method of specifying a finite-state machine is through the use of a transition matrix in which the rows are associated with the present states and the columns are associated with the next states. Entries in the transition matrix represent the inputs and outputs, with a null entry indicating the absence of a transition denoted by a bar.

Example: The transition matrices for machines *M*1 and *M*2 are shown in Fig. 3.4(a) and (b) respectively.

Another method of specifying a finite-state sequential machine is through the use of Algorithmic State Machine (ASM) charts [4]. An ASM

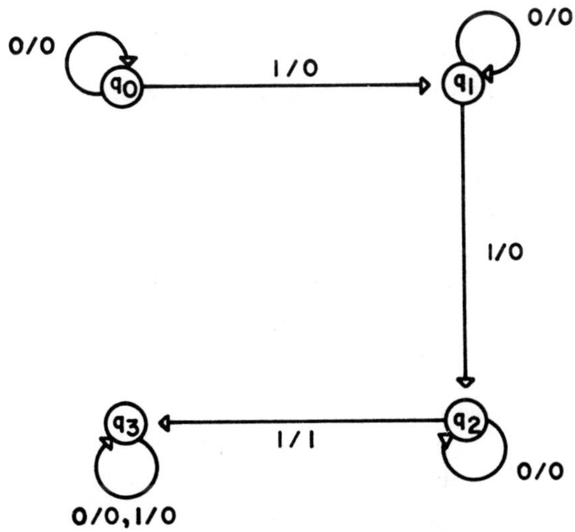

Figure 3.2 State diagram for *M*1

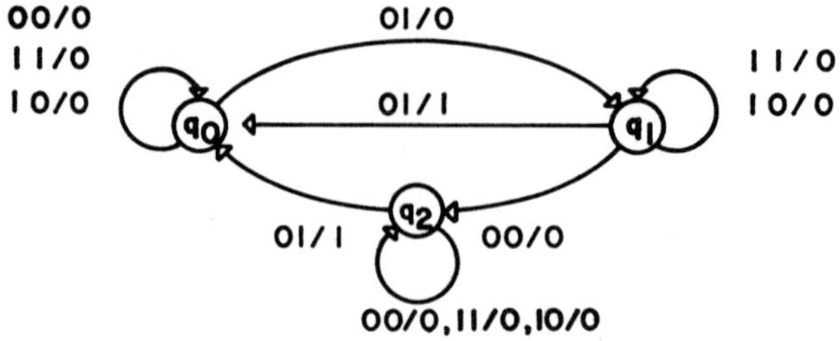

Figure 3.3 State diagram for *M*2

chart resembles a program flowchart and depicts the various state transitions and outputs, conditional and unconditional.

Now let us look more closely at *M*2 whose state diagram is shown in Fig. 3.3. It is not difficult to see what the machine does. The machine simply generates an output of 1 every time a second input of 01 is encountered. If we modify the machine slightly by allowing it to remain at state q_1 instead of going into state q_2 on input of 00 from state q_1, we find that as far as the input and output behaviors are concerned, the original and new machines are identical. This is because once the original machine has entered either q_1 or q_2, the outputs are going to be identical. This prompts us to establish the concept of state equivalence which is essential to the design of optimal sequential circuits. First of all, we make the following definitions.

An input sequence (input tape or input string) is a sequence of input symbols $I^1I^2 \ldots I^i \ldots I^k$ applied as inputs where I^i is the ith input and k, which is the number of symbols in the sequence, is the length of the

	q_0	q_1	q_2	q_3
q_0	0/0	1/0	—	—
q_1	—	0/0	1/0	—
q_2	—	—	0/0	1/1
q_3	—	—	—	0/0,1/0

(a) machine *M*1

	q_0	q_1	q_2
q_0	00/0,11/0,10/0	01/0	—
q_1	01/1	11/0,10/0	00/0
q_2	01/1	—	11/0,10/0,00/0

(b) machine *M*2

Figure 3.4 Transition matrices

sequence. An output sequence is similarly a sequence of output symbols $Z^1Z^2...Z^i...Z^k$ produced as outputs. The symbol J will be used to denote an input sequence of arbitrary length, and the symbol J_k will be used to denote an input sequence of length k.

It is now possible to extend the definitions of the next-state and output functions to input sequences where δ is a mapping from $\{J_k \times Q\}$ into Q and w is a mapping from $\{J_k \times Q\}$ into the set of output sequences. The function w will now be known as the output-sequence function.

Example: For machine $M1$

$$\delta(0011,q_1) = q_3$$

$$\delta(01010,q_0) = q_2$$

$$w(0011,q_1) = 0001$$

$$w(01010,q_0) = 00000$$

Let q_i and q_j be in Q. Then q_i and q_j are *equivalent*, denoted as $q_i \equiv q_j$ if and only if $w(J_k,q_i) = w(J_k,q_j)$ for all possible J_k. A weaker equivalence property can be defined and it is the k-equivalence. Two states q_i and q_j are said to be k-equivalent, denoted as $q_i \overset{k}{=} q_j$ if and only if $w(J,q_i) = w(J,q_j)$ for all possible J of length less than or equal to k.

It can be shown that both the equivalence and the k-equivalence operations defined as above are equivalence relations obeying the reflexive, symmetric, and transitive laws. Each equivalence relation can therefore be used to partition Q, the set of states, into equivalence classes. We shall use the notations π and π_k to denote partitions of Q obtained from the equivalence and the k-equivalence operations respectively.

Let $\pi = \{B_1,B_2,...,B_m\}$ where B_i is called a block of π and is a set of equivalent states and m is the number of equivalent classes and hence is the number of blocks in π.

Since two states in a block are equivalent to each other, if we are interested only in the external behavior of the machine, it will not matter which of the two states the machine starts from on application of a certain input sequence; identical output sequences will be produced. Thus one of the states is redundant and can be removed from Q as long as any occurrence of that state in a description of the machine is replaced by another equivalent state. This means that it is only necessary to retain one state in each block in π to define a new machine whose external behavior is identical to that of the original machine. In this new machine, Q will contain m elements. In addition, there will exist no other machine whose state set Q has fewer than m elements and whose external behavior is identical to that of the original machine. A machine defined in this way is minimal (optimal) since it has the fewest possible number of states.

It will be seen later in the chapter that the number of states in a machine directly affects the amount of storage required to realize the machine as a sequential circuit (the number of memory units required reduces by one if the number of states is reduced from $2^{n-1} + 1$ to 2^{n-1} where n is an integer) and hence the concept of an optimal machine is important in the design of sequential circuits. Thus in general, the fewer states there are, the simpler the implementation would be for the machine. We now briefly describe a procedure which can be used to obtain a minimal machine for a given (completely specified) machine. The procedure will be given through the use of an example as illustration.

Consider machine $M3$ whose state table is given in Table 3.3.

Step 1: A partition π_1 is obtained in which states are grouped in the same block if they are 1-equivalent. Two states q_i and q_j are in the same block in π_1 if and only if $w(J_1,q_i) = w(J_1,q_j)$ for all possible J_1. By inspection $M3$ can be rearranged as shown in Table 3.4 yielding

$$\pi_1 = \{\overline{q_0,q_2,q_4};\ \overline{q_1,q_3,q_5}\}$$

Step 2: A π_1 table is formed in which states are grouped according to blocks, and each entry in the table is the next state tagged with the block number in which the state occurs. The π_1 table for the example is shown in Table 3.4.

Present state	Next state 0	Next state 1	Output 0	Output 1
q_0	q_1	q_1	1	0
q_1	q_0	q_3	0	1
q_2	q_1	q_1	1	0
q_3	q_2	q_1	0	1
q_4	q_5	q_3	1	0
q_5	q_2	q_4	0	1

Table 3.3 State table for $M3$

B_i	Present state	Next state 0	Next state 1
B_1	q_0	$q_1,2$	$q_1,2$
	q_2	$q_1,2$	$q_1,2$
	q_4	$q_5,2$	$q_3,2$
B_2	q_1	$q_0,1$	$q_3,2$
	q_3	$q_2,1$	$q_1,2$
	q_5	$q_2,1$	$q_4,1$

Table 3.4 π_1 table for $M3$

Step 3: Given π_k, an attempt is now made to obtain a finer partition π_{k+1}. From the information contained in the π_k table, two states q_i and q_j are placed in the same block in π_{k+1} if and only if they are in the same block of π_k and $\delta(J_1,q_i) \equiv \delta(J_1,q_j)$ for all possible J_1. If a finer partition is obtained, we go to Step 4. On the other hand, if a finer partition is not possible, then the most recent partition is π and we go to Step 5. For M3,

$$\pi_2 = \{\overline{q_0,q_2,q_4};\ \overline{q_1,q_3};\ \overline{q_5}\},$$

$$\pi_3 = \pi_4 = \{\overline{q_0,q_2};\ \overline{q_4};\ \overline{q_1,q_3};\ \overline{q_5}\}.$$

Step 4: The partition table is formed in a manner as in Step 2. Now we go back to Step 3. The π_2 table and π_3 table are shown in Tables 3.5 and 3.6.

B_i	Present state	Next state 0	Next state 1
B_1	q_0	$q_1,2$	$q_1,2$
	q_2	$q_1,2$	$q_1,2$
	q_4	$q_5,3$	$q_3,2$
B_2	q_1	$q_0,1$	$q_3,2$
	q_3	$q_2,1$	$q_1,2$
B_3	q_5	$q_2,1$	$q_4,1$

Table 3.5 π_2 table for $M3$

B_i	Present state	Next state 0	1
B_1	q_0	q_1,2	q_1,2
	q_2	q_1,2	q_1,2
B_4	q_4	q_5,3	q_3,2
B_2	q_1	q_0,1	q_3,2
	q_3	q_2,1	q_1,2
B_3	q_5	q_2,1	q_4,4

Table 3.6 π_3 table for $M3$

Step 5: A state table for the minimal machine defined by π can be obtained from the most recent partition table by arbitrarily retaining one of the states in each block, replacing each of the rest of the states in a block by the state retained for that block, deleting identical rows, and finally inserting back an appropriate column for outputs. Table 3.7 shows the state table of a minimal machine for the given machine $M3$; note that Q for the optimal machine now contains only 4 elements.

The above procedure is based on the facts that if $q_i \overset{k}{\equiv} q_j$ and $\delta(J_1,q_i) \overset{k}{\equiv} \delta(J_1,q_j)$ then $q_i \overset{k+1}{\equiv} q_j$ and that $\pi = \pi_k$ when $\pi_k = \pi_{k+1}$ [2, 5,6]. It can also be shown that if Q contains p elements, then the partition process in Step 3 will terminate for $k \leq p - 1$; this means that the above procedure will always terminate in a finite number of steps.

Present state	Next state 0	1	Output 0	1
q_0	q_1	q_1	1	0
q_4	q_5	q_1	1	0
q_1	q_0	q_1	0	1
q_5	q_0	q_4	0	1

Table 3.7 State table of minimal machine for $M3$

3.2 SEQUENTIAL CIRCUITS

The storage or memory unit is essential in sequential circuits. It is possible to design sequential circuits in which memory units are not present as individual elements, but are implemented through use of the inherent delay properties of the associated logic gates; such a memory unit can be looked upon as a delay element as shown in Fig. 3.5(a) where d is the delay introduced by the delay element D. Thus, in a D- element, the input signal appears at the output terminal after a period equal to the delay d of the element.

Other commonly used memory elements called flip-flops include the J-K flip-flop, the R-S flip-flop, toggle T flip-flop, and a few other more complex ones. The J-K, R-S and T flip-flops together with their terminal characteristics are shown in Fig. 3.5(b), (c) and (d). As can be seen from the characteristics shown in Fig. 3.5, a flip-flop is a bistable (two-state) element capable of producing an output of one of two values depending on the inputs and the previous output. In a T flip-flop, the value of Q is reversed each time a 1 is applied to x. In the J-K and R-S flip-flops, setting J and S to a 1 makes Q a 1 and setting K and R to a 1 makes Q a 0. In the case of an R-S flip-flop, the input combination 1,1 for $R(t)$, $S(t)$ should be avoided to produce determinate results.

Flip-flops can be implemented using logic gates with feedback. As examples, an R-S flip-flop and a J-K flip-flop implemented using NAND gates are shown in Fig. 3.6. Note that for this particular J-K flip-flop, in addition to the $J(t)$ and $K(t)$ inputs, there is a $C(t)$ input from a clock which is used to control the operation of the circuit so that changes are allowed only at fixed intervals; circuits of this kind are known as synchronous sequential circuits in contrast to unclocked or asynchronous ones. The simple R-S flip-flop shown in Fig. 3.6(a) is also known as a latch.

In many circuits where flip-flops are used, it is often required to simultaneously read out the present content of a flip-flop and store new information into it. It is not difficult to visualize the kind of problem that will be encountered if the flip-flop is of a kind as shown in Fig. 3.6(b); if the clock period is sufficiently long, the new content will be read out as well, producing erroneous results. The clock period, on the other hand, cannot be reduced beyond limit which is the minimum amount of time required to trigger a state transition in the flip-flop. One way to eliminate the problem is to use a master-slave flip-flop, which is formed through the concatenation of two ordinary flip-flops or latches. An R-S master-

$$Q(t) = x(t - d)$$

(a) D-element

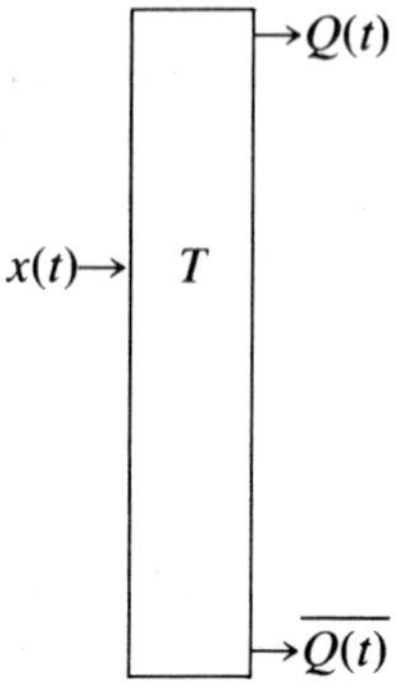

$x(t-1)$	0	0	1	1
$Q(t-1)$	0	1	0	1
$Q(t)$	0	1	1	0

(b) *T* flip-flop

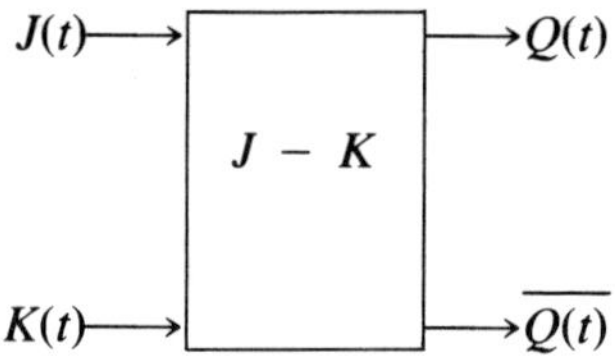

$Q(t-1)$	0	0	0	0	1	1	1	1
$J(t-1)$	0	0	1	1	0	0	1	1
$K(t-1)$	0	1	0	1	0	1	0	1
$Q(t-1)$	0	0	1	1	1	0	1	0

(c) $J - K$ Flip-flop

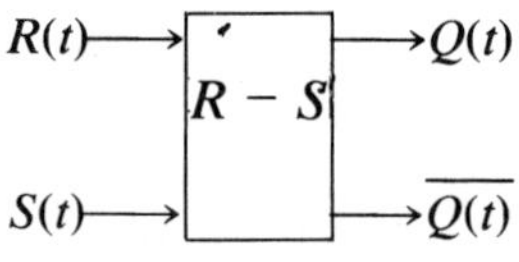

$Q(t-1)$	0	0	0	0	1	1	1	1
$R(t-1)$	0	0	1	1	0	0	1	1
$S(t-1)$	0	1	0	1	0	1	0	1
$Q(t)$	0	1	0	id*	1	1	0	id*

* id denotes indeterminate

(d) $R - S$ flip-flop

Figure 3.5 Storage elements

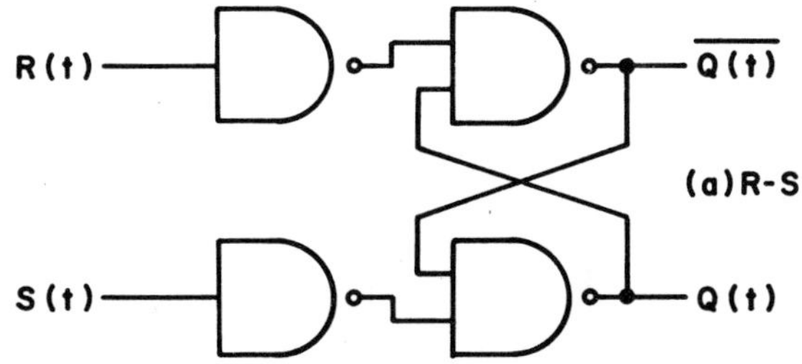

(a) R-S flip-flop

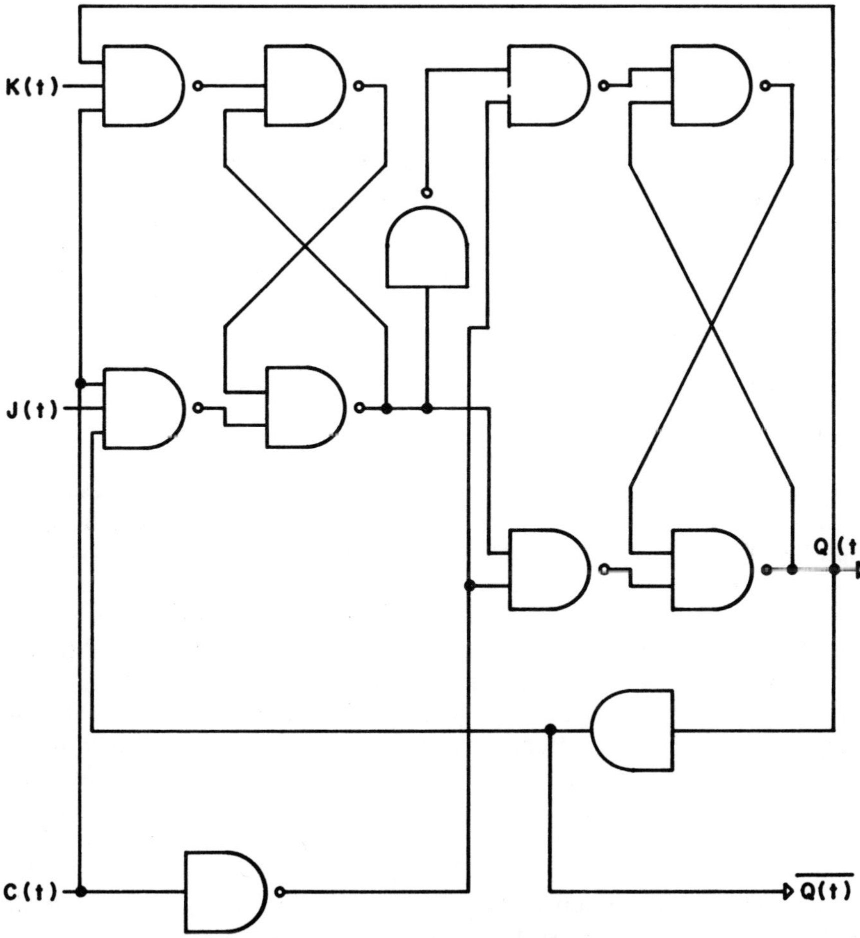

(b) J-K flip-flop

Figure 3.6 Flip-flop implementation using NAND gates

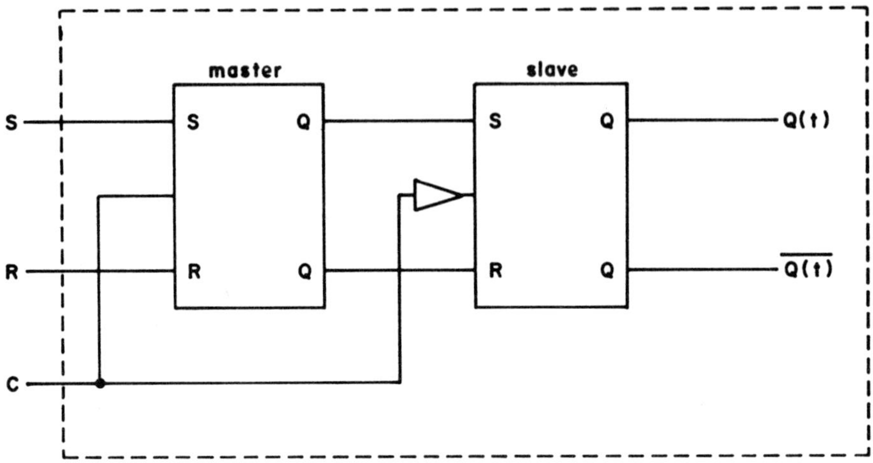

Figure 3.7 A master-slave flip-flop

slave flip-flop is shown in Fig. 3.7. The input signal goes into the master flip-flop and the output is taken from the slave flip-flop. The content in the master flip-flop is transferred into the slave flip-flop a short time after the clock pulse has subsided, ensuring the integrity of the output read out during the high period of the clock pulse. Master-slave flip-flops are very widely used in circuit design.

We can, as indicated in the previous paragraph, differentiate between two types of sequential circuits, the synchronous or clocked sequential circuits and the asynchronous or unclocked sequential circuits. In more complex systems, it is possible to find a number of synchronous subsystems interconnected to form one asynchronous system. Higher operational speeds can usually be achieved with asynchronous circuits; but as will be seen in later, asynchronous circuits are generally more difficult to design due mainly to timing complications.

The general configuration of a synchronous sequential circuit is shown in Fig. 3.8(a). The combinational circuit has $m + k$ input lines. The m-tuple $(x_1, x_2, \ldots, x_m)$ is the primary input and the k-tuple $(y_1, y_2, \ldots, y_k)$ denotes the set of values stored in the memory unit which represents the state of the sequential circuit. Each y_i, $i = 1$ to k, is called a state variable. The memory can be a collection of k delay elements or flip-flops of one of the various types as discussed earlier. The clock is a pulse generator; both the frequency and the pulse width used have to be chosen carefully in order to result in proper operation of the sequential circuit. There are $n + k$ output lines from the combinational circuit. The n-tuple $(z_1, z_2, \ldots, z_n)$

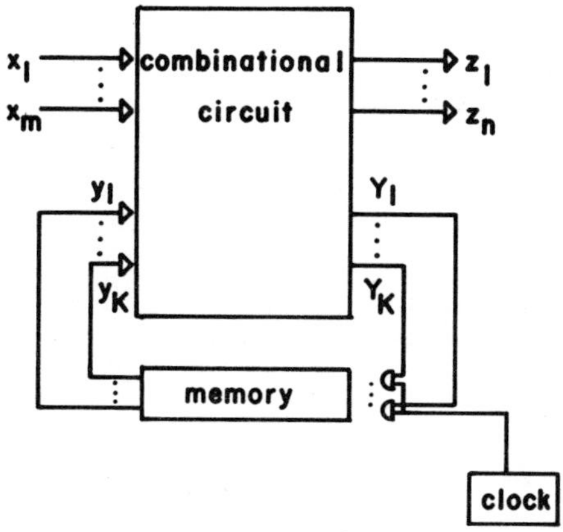

Figure 3.8 (a) General configuration of a synchronous sequential circuit

denotes the output, and the k-tuple $(Y_1, Y_2, \ldots, Y_k)$ represents the next state of the sequential circuit.

The configuration of an asynchronous sequential circuit is shown in Fig. 3.8(b). It is essentially the same as the synchronous sequential circuit except for the absence of a clock. The memory is either made up of explicit delay elements or is simply implied by the inherent delays of the elements in the combinational circuit. Changes of the state in an asynchronous circuit are governed by changes in the inputs in a manner dependent on the various delays within the circuit. Timing considerations are especially critical in the design of asynchronous sequential circuits.

Sequential circuits are realizations of finite-state machines. For a given finite-state machine, we wish to develop efficient procedures to obtain "good" sequential circuits to realize the machine. The term "good" of

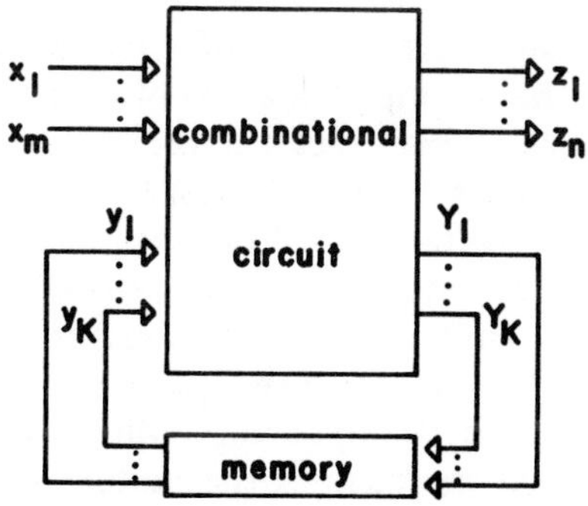

Figure 3.8 (b) General configuration of an asynchronous sequential circuit

course requires a more specific definition which depends on operational requirements. Some specific criteria may include the number of gates in the combinational subcircuit and the number of memory elements. In general, a design procedure aims at obtaining a sequential circuit which requires the least amount of memory, although this is becoming less and less true as memory becomes cheaper and cheaper. We will discuss this further in Section 3.3.

The design of synchronous sequential circuits will be discussed first, and the design of asynchronous sequential circuits will be picked up later in Section 3.5.

3.3 DESIGN OF SYNCHRONOUS SEQUENTIAL CIRCUITS

The design of a synchronous sequential circuit can normally be divided into a number of steps. The first step involves the transformation of the description of the circuit to be designed into the form of a state diagram. In the second step, the state diagram is translated into a state table representation of a finite-state machine. Depending on requirements, an optimization step is then taken to obtain the state table of an optimal machine using the method discussed in Section 3.1. The desired sequential circuit is then realized in the final step of the design procedure. In certain cases, it may be possible to establish a state table directly from the circuit description without having to obtain a state diagram. Sometimes the circuit can be more efficiently generated directly from the word description without first generating a state table.

We use two running examples in this section to illustrate the operation of the design procedure. One of the examples is the design of a pulse counter that is capable of repeatedly counting from 0 to 10 the number of input pulses, issuing an output of 1 on the count of 10. The other example is the design of a circuit which can be used to detect the occurrence of a serial input bit pattern of 101; an output of 1 is issued whenever it is found that an input of a 1 is preceded by a 0 which in turn is preceded by another 1. We will refer to the two examples as the counter design and the pattern-detector design respectively.

To establish a state diagram, the first thing needed to be done is the identification of a starting state, the state in which the machine can be assumed to be initially. In the case of the counter design, a starting state can easily be taken to be one when the counter has just made an output of 1, i.e., has just counted 10 consecutive pulses and is ready to repeat the counting process. In the case of the pattern-detector design, it becomes necessary to introduce an arbitrary state as the starting state be-

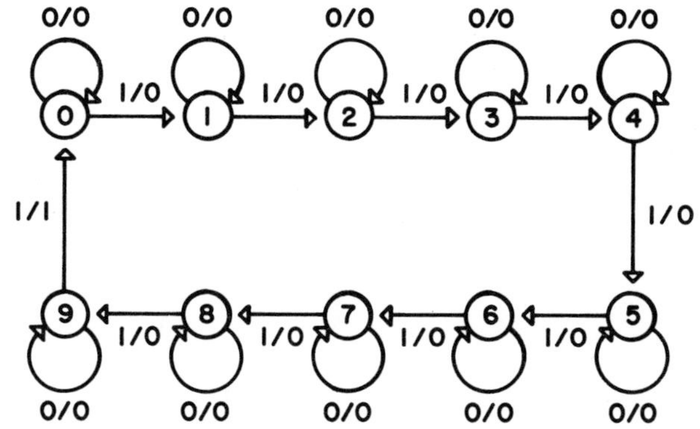

Figure 3.9 State diagram for counter

cause it is impossible to identify from the circuit description, as is possible in the case of the counter design, a point where the pattern detection operation ends and then begins. However, this difficulty does not really bring about any problem in the establishment of the state diagram. Having identified the starting state, it is now (usually) a straight-forward matter to complete the state diagrams. The two state diagrams for the examples are shown in Fig. 3.9 and Fig. 3.10.

In the state diagrams for the counter, a state is labelled by the number of pulses counted. After a count of 10, state 0 is reentered and the counting

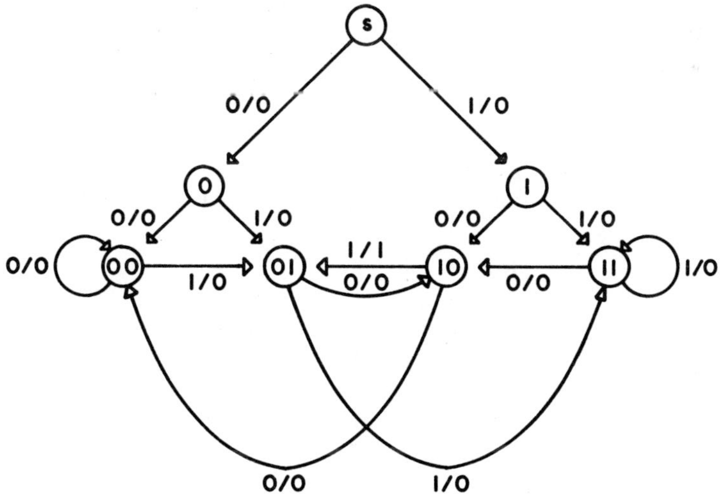

Figure 3.10 State diagram for pattern dectector

Present state	Next state 0	1	Output 0	1
0	0	1	0	0
1	1	2	0	0
2	2	3	0	0
3	3	4	0	0
4	4	5	0	0
5	5	6	0	0
6	6	7	0	0
7	7	8	0	0
8	8	9	0	0
9	9	0	0	1

Table 3.8 State table for counter

process is repeated. A higher-numbered state is entered whenever a pulse is received. The state table for this example is shown in Table 3.8.

In the state diagram for the pattern detector, the starting state is labelled s and all the rest of the states are labelled by the two most recent bits received. Thus state 01 represents the condition when the last two bits are 1 and 0 (the 1 following a 0), and, therefore, if now a 0 comes in, the 3-bit pattern would have been 010 and so a 0 is produced as output; the two most recent bits now becomes 10 and will be used in combination with the next input bit to decide the next correct output and so effectively from state 01, an input of 0 brings the machine to state 10 with an output of 0 and this is indicated by the presence of an arc going from state 01

Present state	Next state 0	1	Output 0	1
s	0	1	0	0
0	00	01	0	0
1	10	11	0	0
00	00	01	0	0
01	10	11	0	0
10	00	01	0	1
11	10	11	0	0

Table 3.9 State table for pattern detector

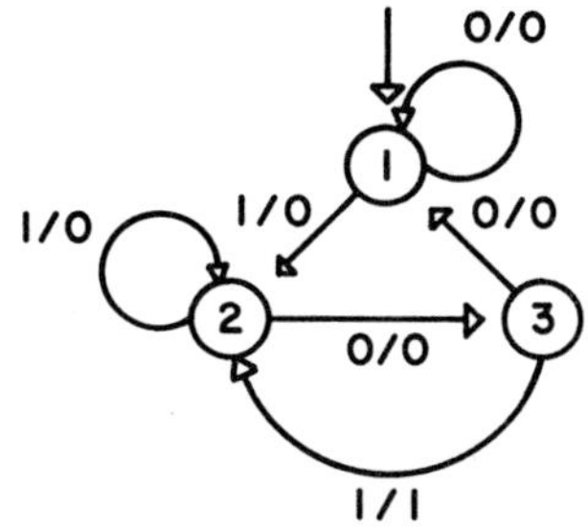

Figure 3.11 State diagram for minimal machine

to state 10 with the label 0/0 on it. The rest of the state diagram can similarly be explained. The state table for this example is shown in Table 3.9.

Having obtained a state table for the machine, it is now possible to apply the method as discussed in Section 3.1 to obtain a minimal-equivalent machine if necessary. It can easily be shown that the machine represented by the state table in Table 3.8 is already a minimal one. On the other hand, we can obtain an equivalent machine with fewer states for the machine shown in Table 3.9; the state diagram and state table for this minimal machine are shown in Fig. 3.11 and Table 3.10 respectively. We have used the labels 1, 2 and 3 for the equivalent classes $\overline{s,0,00}$, $\overline{1,01,11}$ and $\overline{10}$.

Our next step is now to realize each of the finite-state machines using memory elements and a combinational circuit. First, we have to decide on what kind of memory elements to use. Let us assume the use of *R-S* flip-flops. The terminal characteristics of an *R-S* flip-flop has been given in Fig. 3.5(d). The realization of the counter circuit will be followed in detail to illustrate this realization step.

There are a total of ten distinct states. Hence four flip-flops are required since four flip-flops can produce a maximum of $2^4 = 16$ distinct states,

Present state	Next state 0	Next state 1	Output 0	Output 1
1	1	2	0	0
2	3	2	0	0
3	1	2	0	1

Table 3.10 State table for minimal machine

State	State variable combination $y_1y_2y_3y_4$
0	0000
1	0001
2	0010
3	0011
4	0100
5	0101
6	0110
7	0111
8	1000
9	1001

Table 3.11 Secondary assignment for counter

whereas three flip-flops are only sufficient for $2^3 = 8$ states. Let the ten states be coded as in Table 3.11, where y_1, y_2, y_3 and y_4 are the state variables representing the values stored in the flip-flops. The coding of the states is generally known as the secondary assignments of the states. Any coding which assigns a unique code to each state can be used.

It can easily be shown that a particular choice of a secondary assignment generally produces a different combinational circuit and so this coding

Present state				Next state								Output	
				0				1				0	1
y_1	y_2	y_3	y_4	Y_1	Y_2	Y_3	Y_4	Y_1	Y_2	Y_3	Y_4		
0	0	0	0	0	0	0	0	0	0	0	1	0	0
0	0	0	1	0	0	0	1	0	0	1	0	0	0
0	0	1	0	0	0	1	0	0	0	1	1	0	0
0	0	1	1	0	0	1	1	0	1	0	0	0	0
0	1	0	0	0	1	0	0	0	1	0	1	0	0
0	1	0	1	0	1	0	1	0	1	1	0	0	0
0	1	1	0	0	1	1	0	0	1	1	1	0	0
0	1	1	1	0	1	1	1	1	0	0	0	0	0
1	0	0	0	1	0	0	0	1	0	0	1	0	0
1	0	0	1	1	0	0	1	0	0	0	0	0	1

Table 3.12 State table for counter

step directly influences the complexity of the resulting combinational circuit. The topic of the secondary assignments is an important and interesting one and has been well investigated. Interested readers are referred to the references given at the end of the chapter. The number of different assignments is too large for exhaustive analysis even for relatively small tables. We shall simply arbitrarily choose a certain secondary assignment and obtain an optimal combinational circuit for that choice.

With the above choice of a secondary assignment, the state table for the counter shown in Table 3.8 can now be redrawn so that the state variables are included in the state table. This new state table is shown in Table 3.12 and indicates the relationship between the present value of the state variables denoted by y_1, y_2, y_3 and y_4 and the next value of the state variables denoted Y_1, Y_2, Y_3 and Y_4. Each flip-flop has two input terminals, the set and the reset terminals S and R respectively, and so we are actually required to find switching expressions for the following functions:

$$R(y_1),\ S(y_1),\ R(y_2),\ S(y_2),\ R(y_3),\ S(y_3),\ R(y_4),\ S(y_4).$$

Karnaugh maps for all eight of these switching functions can readily be obtained from the state table in Table 3.12. In doing this, it is essential to refer back to Fig. 3.5 to see how the R and S terminals are used to change the old value stored in the flip-flop to the new value. For example, the entries in the Karnaugh map for $S(y_1)$ in Fig. 3.12 are obtained by noting all those instances for which $Y_1 = 1$ and $y_1 = 0$, since for each of these instances (combination of present state and input), $S(y_1)$ has to be 1, and all those instances for which $Y_1 = 1$ and $y_1 = 1$ since for each of these instances, $S(y_1)$ has to be a don't-care. Don't-care entries are also made for present states of 1010, 1011, 1100, 1101, 1110 and 1111. The rest of the entries are 0s. The entire Karnaugh map is shown in Fig. 3.12. The establishment of the other seven Karnaugh maps can similarly be made.

Since indeterminate values will be produced for the cases when $R = S$ in an R-S flip-flop, we can restrict one of the R and S terminals to be the complement of the other, i.e., $S = R'$ and we now need to find the switching expressions for only four functions (Note that an R-S FF with $S = R'$ works just like a clocked delay element). In this case, the expression for each of the R signals is obtained by noting all those instances for each of which the corresponding Y signal has a value of 0. Note that, unlike the previous case, an instance for which $Y = 0$ and $y = 0$ now requires

xy_1 \ $y_2y_3y_4$	000	001	011	010	110	111	101	100
00	0	0	0	0	0	0	0	0
01	x	x	x	x	x	x	x	x
11	x	0	x	x	x	x	x	x
10	0	0	0	0	0	1	0	0

Figure 3.12 Karnaugh map for $S(y_1)$

R to be 1 (i.e., $S = 0$ since $S = R'$) and not a don't care. Don't-cares are only for the present states of 1010, 1011, 1100, 1101, 1100 and 1111.

From the Karnaugh maps for $R(y_1)$, $R(y_2)$, $R(y_3)$ and $R(y_4)$, we obtain the following switching expressions for the functions:

$$R(y_1) = x'y_1' + y_1'y_4' + xy_3'y_4 + xy_1'y_2'$$

$$R(y_2) = x'y_2' + y_2'y_3' + y_2'y_3y_4' + xy_2y_3y_4$$

$$R(y_3) = x'y_3' + xy_1 + y_3'y_4' + xy_3y_4$$

$$R(y_4) = x'y_4' + xy_4$$

$$\text{output } z = xy_1y_4$$

The S signals are simply the respective complements of the R signals. The implementation of these expressions using logic gates will be left to the reader.

The reader can extend these concepts to J-K and T flip-flops.

In the case of the pattern detector design, only two flip-flops are required, and we can employ a similar procedure to arrive at the following expressions for the combinational circuit using the secondary assignment shown in Table 3.13:

$$R(y_1) = x + y_1 + y_2'$$

$$S(y_1) = x'y_1'y_2$$

$$R(y_2) = x'y_2' + x'y_1$$

$$S(y_2) = x + y_1'y_2$$

$$\text{Output } z = xy_1$$

State	Secondary assignment y_1	y_2
1	0	0
2	0	1
3	1	1

Table 3.13 Secondary assignment for detector

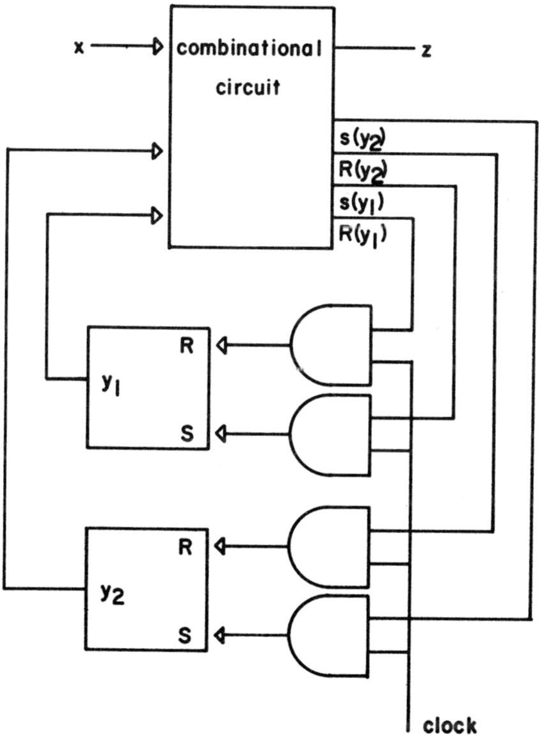

Figure 3.13 Block diagram of pattern detector

In each of the actual circuits realizing the above two examples, a clock input will be required since the circuits are assumed to be synchronous-sequential circuits. For example, the pattern detector circuit will be as shown in Fig. 3.13 in which the combinational circuit defined by the corresponding set of switching expressions is shown as a black box. The flip-flops may actually be master-slave devices.

3.4 INCOMPLETELY SPECIFIED MACHINES

The design procedures discussed in Section 3.3 apply to completely specified machines for which the next-state mapping and the output mapping w are defined for every element in $I \times Q$. Let us look at the incompletely specified machine $M4$ represented by the state table shown in Table 3.14. The blanks in the state table represent don't-care conditions. These don't-care conditions arise in the following way: if from a present state a certain input symbol will never be encountered, or (only for the output entries) if we do not care what the output will be for a particular input, then the corresponding entry will be a don't-care. We can arbitrarily assign values to these don't-care entries thus transforming the original machine into a completely specified one and apply the usual method to obtain an optimal machine. However, the resulting machine may not truly be optimal in the sense that by making another assignment of the don't-care conditions, it may be possible to obtain a machine having even fewer states. Thus a new procedure has to be developed. We shall illustrate the operation of the procedure through an example in which an optimal machine for machine $M4$ is being developed. The reader is referred to the references at the end of this chapter for a more detailed discussion on the subject.

The procedure is based on the mathematical theory of compatible or possible equivalence classes. Two states s_i and s_j in a machine are said

Present state	Next state 0	Next state 1	Output 0	Output 1
q_0	q_1	q_4	0	1
q_1	—	q_5	—	1
q_2	q_4	q_1	1	1
q_3	q_2	q_4	—	1
q_4	q_0	q_5	1	—
q_5	q_2	—	—	—

Table 3.14 State table for $M4$

to be compatible, denoted as $s_i \sim s_j$ if and only if $w(J_1,q_i) = w(J_1,q_j)$ and $\delta(J_1,q_i) \sim \delta(J_1,q_j)$ for all J_1 and all cases where both sides of the equation are specified. Note that compatibility is not transitive and hence is not an equivalence relation. In most cases, the number of compatibles defined this way is very large; and if the states of an optimal machine are to be obtained directly from the set of compatibles, the selection process, which is enumerative in nature, would very often be tedious. We can reduce the number of compatibles that need to be considered in the selection process by defining maximal compatibles and then prime compatibles. The states of an optimal machine are then selected from the set of prime compatibles under the conditions of covering, closure, and minimality [2,5,6,7,8].

Step 1: The first step is to find the set of maximal compatibles. A maximal compatible is a compatible that is not contained in another compatible. To do this, charts are drawn as in Fig. 3.14. The chart in Fig. 3.14(a) shows the implications for the two states to be compatible. For example, for the two states q_3 and q_2 to be compatible, their next states for input 0, namely, q_2 and q_4 respectively and those for input 1, namely, q_4 and q_1 respectively have to be compatible too; i.e., we require $q_2 \sim q_4$ and $q_4 \sim q_1$, an implication indicated by the entries q_2q_4 and q_4q_1 in the appropriate cell in the chart. Also, since $w(0,q_0) \neq w(0,q_2)$, q_0 and q_2 can never be compatible and so a cross is entered. No implication is required for states q_5 and q_3 to be compatible and so a checkmark is entered. Starting from the rightmost column in chart (a) and moving left, those cells with single crosses as entries are used to insert single crosses into appropriate cells to indicate that the corresponding states cannot be compatible; a second cross is entered into each of the former cells. This testing process is continued until no cells remain in which there is only a single cross. For the example, the end result is shown in Fig. 3.14(b). What remains now is to identify from Fig. 3.14(b) the set of maximal compatibles. This is done in Fig. 3.15. Starting with the rightmost column, we see that q_4 and q_5 are not compatible and so they are placed in two separate classes. Moving on to column q_3, we find that q_3 is compatible with q_5 and so q_3 and q_5 now forms a new class. Next q_2 is inserted as a new class since it is not compatible with q_3, q_4 or q_5. The state q_1 is compatible with all of q_2, q_4 and q_5 but not with q_3; so three new classes q_1q_2, q_1q_4 and q_1q_5 are formed. Finally q_0

q_1	q_4q_5				
q_2	X	q_1q_5			
q_3	q_1q_2	q_4q_5	q_2q_4 q_1q_4		
q_4	X	√	q_0q_4 q_1q_5	q_0q_2 q_4q_5	
q_5	q_1q_2	√	q_2q_4	√	q_0q_2
	q_0	q_1	q_2	q_3	q_4

(a)

q_1	q_4q_5 X				
q_2	XX	q_1q_5			
q_3	q_1q_2	q_4q_5 X	q_2q_4 X q_1q_4		
q_4	XX	√	q_0q_4 X q_1q_5	q q q q	
q_5	q_1q_2	√	q_2q_4 X	√	q_0q_2 X
	q_0	q_1	q_2	q_3	q_4

(b)

Figure 3.14 Charts to identify maximal compatibles

q_4: $\overline{q_4}$; $\overline{q_5}$

q_3: $\overline{q_4}$; $\overline{q_3, q_5}$

q_2: $\overline{q_2}$; $\overline{q_4}$; $\overline{q_3, q_5}$

q_1: $\overline{q_1, q_2}$; $\overline{q_1, q_4}$; $\overline{q_3, q_5}$; $\overline{q_1, q_5}$

q_0: $\overline{q_1, q_2}$; $\overline{q_1, q_4}$; $\overline{q_0, q_3, q_5}$; $\overline{q_1, q_5}$

Figure 3.15 Finding maximal compatibles

is included in the class formed by q_3 and q_5. The set of maximal compatibles are thus

$$\{\overline{q_1,q_2};\ \overline{q_1,q_4};\ \overline{q_0,q_3,q_5};\ \overline{q_1,q_5}\}$$

Step 2: The next step is to identify from the set of maximal compatibles the set of prime compatibles. We shall only attempt an informal definition of a prime compatible. First, we define the closure set of a compatible to be the set of compatibles other than singletons, implied by and not included in the given compatible. In the given example, the closure set of the compatible $\overline{q_2,q_3}$ is $\{\overline{q_1,q_4},\ \overline{q_2,q_4}\}$ as can be seen from Fig. 3.14(b). Second, a compatible c_j is excluded by a compatible c_i if $c_i \supsetneq c_j$ and the closure set of c_i is included in the closure set of c_j. Finally, a compatible is said to be prime if it is not excluded by any other compatible. What is needed now is to obtain the set of prime compatibles.

Example: In the example, neither of the compatibles $\overline{q_0,q_3}$, $\overline{q_0,q_5}$, is prime since the closure sets of $\overline{q_0,q_3,q_5}$, $\overline{q_0,q_3}$, and $\overline{q_0,q_5}$ are all $\{\overline{q_1,q_2}\}$. On the other hand $\overline{q_3,q_5}$ is a prime. The set of prime compatibles are enumerated from the set of maximal compatibles, shown for this example in Table 3.15. Note that we start with the maximal compatibles having the most number of states. The other compatibles and new ones derived from them are added successively. A cross is inserted beside a compatible whenever it is found to be excluded by any other compatible. Note that a maximal compatible is also a prime compatible but that the reverse is not always true. For the example, the final list of prime compatibles is then

$$\{\overline{q_0},\ \overline{q_1},\ \overline{q_2,q_4},\ \overline{q_1,q_2},\ \overline{q_1,q_4},\ \overline{q_1,q_5},\ \overline{q_0,q_3,q_5}\}$$

Step 3: The selection from the final list of a minimal closed cover of prime compatibles is a very difficult problem and the usual procedure is based on the fact that an optimal machine can be obtained by considering only the set of prime compatibles. For the example, under the required conditions of covering, closure, and minimality, the following set of compatibles is chosen and can be shown to be minimal:

$$\{\overline{q_0},\ \overline{q_2},\ \overline{q_1,q_4},\ \overline{q_3,q_5}\}$$

Number of states	Compatible	Closure set
$n = 3$	$\overline{q_0,q_3,q_5}$	$\{\overline{q_1,q_2}\}$
$n = 2$	$\times$ $\overline{q_0,q_3}$	$\{\overline{q_1,q_2}\}$
	$\times$ $\overline{q_0,q_5}$	$\{\overline{q_1,q_2}\}$
	$\overline{q_3,q_5}$	$\{\phi\}$
	$\overline{q_1,q_2}$	$\{\overline{q_1,q_5}\}$
	$\overline{q_1,q_4}$	$\{\phi\}$
	$\overline{q_1,q_5}$	$\{\phi\}$
$n = 1$	$\overline{q_0}$	$\{\phi\}$
	$\overline{q_1}$	$\{\phi\}$
	$\overline{q_2}$	$\{\phi\}$
	$\times$ $\overline{q_3}$	$\{\phi\}$
	$\overline{q_4}$	$\{\phi\}$
	$\times$ $\overline{q_5}$	$\{\phi\}$

Table 3.15 Finding prime compatibles

Note that we could also have chosen the set

$$\{\overline{q_4}, \overline{q_1,q_2}, \overline{q_1,q_5}, \overline{q_0,q_3,q_5}\}$$

and a machine with the same number of states is obtained.

Step 4: Using the chosen set of compatibles, the optimal machine is obtained in a manner similar to that for the completely specified machine case. For the example, the optimal machine is shown in Table 3.16.

Present state	Next state		Output	
c_0	c_2	c_2	0	1
c_1	c_2	c_2	1	1
c_2	c_0	c_3	1	1
c_3	c_1	c_2	—	1

Table 3.16 Optimal machine for *M*4

3.5 DESIGN OF ASYNCHRONOUS SEQUENTIAL CIRCUITS

The design of an asynchronous sequential circuit is similar to that of a synchronous sequential circuit discussed in the previous sections. However, as we shall shortly see, several additional complications arise which render the design procedure of asynchronous sequential circuit much more involved.

Let us reexamine the design of the pattern detector circuit discussed in Section 3.3. This time we shall design it as an asynchronous sequential circuit instead. The state table for the optimal machine shown in Table 3.10 is redrawn in Table 3.17. This state table will be used to demonstrate the special considerations which need to be taken in order to be able to arrive at any asynchronous circuit which works correctly.

Suppose the machine is in state 1. As long as the input remains at 0, the machine will remain in state 1. Now suppose the input is changed to 1. From the state table in Table 3.17, it can be seen that the state will change from state 1 to state 2 and as long as the input remains at 1, the machine will remain in state 2. Now suppose the input is changed back to 0. The correct state to be entered and maintained should be state 3 but examination of the state table indicates that state 3 will indeed be entered but that state 1 will next be entered and maintained as long as the input remains at 0. State 3 is only a transient state. Thus, the state table shown in Table 3.17 does not correctly describe the asynchronous detector circuit.

Let us examine the behavior of the machine in more detail. Starting from the very beginning, the input x is 0 and we can designate this condition as state 1. Now suppose x is changed to 1. State 1 now goes to another state which will be called state 2. It is represented as row 1 in Table 3.18 which is known as a *flow table*. Each entry in the flow table consists of a state and an output. We distinguish between two types of states, stable ones and unstable ones. A state is stable if it is maintained as long as the input remains unchanged; otherwise it is unstable. In a flow table, a stable state is enclosed with a circle. Thus in row 1 in the flow

Present state	Next state 0	Next state 1	Output 0	Output 1
1	1	2	0	0
2	3	2	0	0
3	1	2	0	1

Table 3.17 State table for optimal machine

state 0	output 1
①,0	2 ,0
3 ,0	②,0
③,0	4 ,1
3 ,0	④,1

Table 3.18 Flow table for pattern detector

table in Table 3.18, state 1 is a stable state and state 2 is not. Continuing with the input sequence, state 2 has to be maintained if the input remains at 1. On the other hand, if the input is now changed to 0, then a new state should be maintained as long as the input remains at 0. This explains the entries in row 2 and also the first entry 3 in row 3 in Table 3.18. If while in state 3, the input is changed from 0 to 1, the input 101 has occurred; so a 1 should be issued as output and a new state, 4, is entered. State 4, then, is the state in which an output of 1 is being produced. It is not difficult to realize that from state 4, on change of input, state 3 should be reentered. We thus have a complete flow table describing the asynchronous detector circuit.

The previous example illustrates the generally used procedure to obtain a flow table from a description of the behavior of an asynchronous circuit to be designed. All possible flows of states have to be considered, new states being added as needed in the process. The resulting asynchronous flow table can be converted into an (usually incompletely specified) asynchronous state table by adding a column to the left hand side of the flow table, each entry in the column being the corresponding stable state in that row. Thus the optimal asynchronous state table derivable from the flow table shown in Table 3.18 is given in Table 3.19 (Note that states 3 and 4 in Table 3.18 are equivalent); in this particular case the state table

Present state	Next state 0	Next state 1	Output 0	Output 1
1	1	2	0	0
2	3	2	0	0
3	3	3	0	1

Table 3.19 State table derived from flow table

is a completely specified one. From this point on, state reduction techniques as described in Sections 3.1 and 3.3 can be applied to arrive at optimal machines. The reader should note that Table 3.19 is not equivalent to the original Table 3.17. This is because an asynchronous state table assumes level alternating inputs while a synchronous state table can have input sequences with repeated inputs (e.g. 0001).

Now let us continue with our design using the state table in Table 3.19. Since we have three distinct states, a minimum of two storage elements are required. Let the following secondary assignment be used:

State	y_1	y_2
1	0	0
2	0	1
3	1	0

and let T flip-flops be used as storage elements. The state table is now redrawn in Table 3.20. using the state variables y_1 and y_2.

Karnaugh maps can be established to arrive at the following expressions for the combinational circuit for the state variables (the actual derivation of the expressions is left as an exercise for the reader):

$$T(y_1) = x'y_2$$

$$T(y_2) = x'y_2 + xy_2'y_1'$$

$$\text{output } z = xy_1y_2'$$

It is now an easy matter to build a circuit based on the illustrated set of switching expressions. Unfortunately, the circuit, if built, will not operate correctly. Let us see why. Suppose the state is stable at $y_1 = 0$ and

Present state		Next state				Output	
		0		1		0	1
y_1	y_2	Y_1	Y_2	Y_1	Y_2		
0	0	0	0	0	1	0	0
0	1	1	0	0	1	0	0
1	0	1	0	1	0	0	1

Table 3.20 State table with state variables

$y_2 = 1$ and input $x = 1$. Now let input x be changed from 1 to 0. The state table indicates that y_1 should change from 0 to 1 and y_2 should change from 1 to 0 and the circuit should remain in that state as long as the input is unchanged. However, because the circuit is operating asynchronously, the changes in y_1 and y_2 will not occur simultaneously. Suppose y_2 changes first. This means that from state 01 state 00 is actually entered and from the state table in Table 3.20 it can be seen that state 00 is stable. So an erroneous operation results since state 10 is never entered. Alternatively, if y_1 changes first, i.e., we have $y_1 = 1$ and $y_2 = 1$, then the previously derived set of equations for $T(y_1)$ and $T(y_2)$ gives $T(y_1) = 1$, $T(y_2) = 1$. Thus state 10 is entered and for this case, the operation is correct.

The above example illustrates what is known as a race problem in asynchronous circuits. A *race* occurs whenever there is a need for a change in more than one of the secondary variables. A race is critical if the final state to be reached depends on the order in which the changes in the secondary variables occur and is noncritical otherwise. In the above example, the race is a critical one.

Because of the presence of a critical race in the earlier designed circuit, we consider an alternative design using the following secondary assignment scheme:

State	y_1	y_2
1	0	0
2	0	1
3	1	1

With this new secondary assignment scheme, the state table is now redrawn in Table 3.21. The corresponding set of switching expressions describing the combinational circuit is then

$$T(y_1) = x'y_1'y_2$$

$$T(y_2) = xy_2'$$

$$\text{output } z = xy_1$$

A circuit based on the above design does not exhibit any race, critical or noncritical. This is due to the fact that during any state transition, one and only one of the state variables changes value. We have actually se-

Present state		Next state				Output	
		0		1		0	1
y_1	y_2	Y_1	Y_2	Y_1	Y_2		
0	0	0	0	0	1	0	0
0	1	1	1	0	1	0	0
1	1	1	1	1	1	0	1

Table 3.21 State table with state variables

lected a secondary assignment scheme in which each pair of states between which there is a transition have been assigned state variable combinations which differ in value in only one of the variables. This can be clearly illustrated by plotting the secondary assignment for the example in a Karnaugh map form as shown in Fig. 3.16. States 1 and 2 and states 2 and 3 occupy adjacent pairwise cell positions.

It is sometimes not possible to find a secondary assignment in this straightforward manner to satisfy the pairwise adjacency condition. However, it is possible to design race-free circuits by increasing the number of state variables and generally the number of storage elements. Systematic procedure for designing race-free circuits can be found in most of the references given at the end of this chapter.

Another kind of phenomenon which can render the operation of an asynchronous circuit incorrect is the occurrence of hazards. A hazard is due to the differing time delays incurred by the different signals coming into the input terminals of a gate; as a result, the gate may not assume the desired value. Thus hazards are related to the combinational circuit part of the sequential circuit. By carefully choosing switching expressions used in the implementation of the combinational circuit, with the implied intention of including redundant terms if necessary, it is also possible to arrive at hazard-free circuits [5,8].

y_1 \ y_2	0	1
0	1	2
1	—	3

Figure 3.16 Karnaugh map showing secondary assignment

3.6 DESIGN EXAMPLES AND PRACTICAL CONSIDERATIONS FOR IMPLEMENTATION

In the previous sections, a sequential circuit was designed using the concept of states. As the circuit to be designed becomes more and more complex, it is evident that this approach becomes less and less feasible because the number of states gets unmanageably larger. Another approach must then be taken. The circuit is decomposed into convenient subcircuits which can of course in turn be decomposed repeatedly. The subcircuits are designed separately and then each subcircuit is generally designed as a synchronous-sequential circuit which either uses a separate clock or shares a clock with one or more subcircuits. It is, of course, necessary to ensure that the final interconnection of the subcircuits indeed results in the desired circuit. The decomposition process is usually carried out in such a way that common-functional subcircuits, like decoders, encoders, latches, and counters are identified, especially when most of these well-recognized subcircuits are now readily available commercially in single-chip integrated circuits. A circuit thus designed can be called a network. We shall illustrate this important design approach through the description of a couple of examples.

First we take up the design of the simplified traffic light controller which has been partially described in Section 2.2. With the introduction of the counters T_A, T_B and T_C, the controller can now be drawn in block diagram form as that shown in Fig. 3.17. Note that for clarity, appropriate clocks have not been shown in the figure. Subcircuit S is a combinational circuit described by the following set of switching expressions (Section 2.2):

$$S_A = c + t_B$$

$$S_B = t_A c'$$

$$S_C = c + t_B$$

$$1_A = c + t_B$$

$$1_B = t_A c'$$

Counter T_C has an additional external pulse input coming from the transducer indicating the oncoming of another car. The realization of subcircuit S using gates is straightforward and will not be pursued.

The three counters T_A, T_B and T_C, are obviously functionally similar. T_C counts the number of external input pulses beginning from 0 and produces an output c when the count reaches 5. S_C resets T_C and the counting

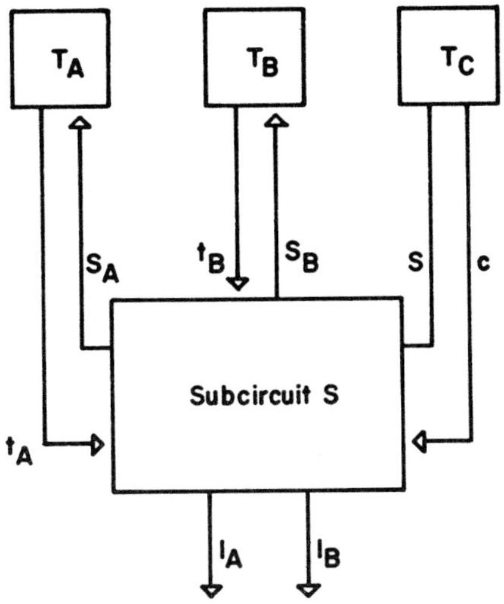

Figure 3.17 Traffic light controller

process repeats itself. T_A is preset by S_A to an initial value and counts down, using the output from a clock with the appropriate frequency, to 0 in 20 seconds at which time an output is produced on t_A. Or equivalently, T_A is made to count up from 0 to a precalculated value in 20 seconds at which time an output is produced on t_A and at the same time T_A is reset to 0 by S_A. The operation of T_B is similar to that of T_A. Integrated counter chips are available commercially at very low costs.

Let us now look at a second example which is based on [3]. The circuit to be designed is a service-request controller. The service request controller is used to connect a number of peripheral devices to a central processing unit. The peripheral devices issue requests for service by the central processing unit and the requests are to be treated on a fixed priority basis. We give a more detailed description of this controller design in the following.

A block diagram of the controller is shown in Fig. 3.18. Service requests from the peripheral devices come in through the input lines $x_0,\ldots,x_{n-1}$. Thus $x_i = 1$ if the device with the assigned address i requests service by the CPU. At any moment, there can be more than one device waiting for service. A device with the lowest address is to be serviced ahead of others in our fixed-priority scheme. The address of the device selected by the controller to be serviced is sent to the CPU through the output lines $y_1,\ldots,y_m$ where it is assumed that $2^{m-1} < n \leq 2^m$. The CPU is informed

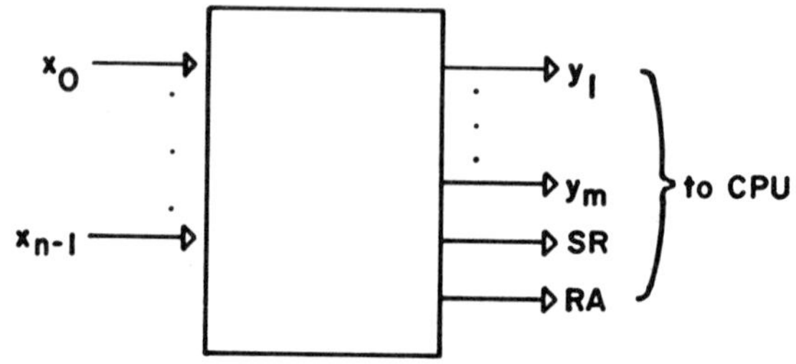

Figure 3.18 Service request controller

of the presence of the request by the output on the line *SR*. Once the CPU has received the *SR* signal, it sends back a signal, in the form of a pulse, through the line *RA* to the controller, indicating that the CPU has acknowledged the request and this signal is then used by the controller to reset *SR* to 0. As soon as service has been offered to the device with address *i*, x_i is reset to 0 (by the device itself or the interface connecting

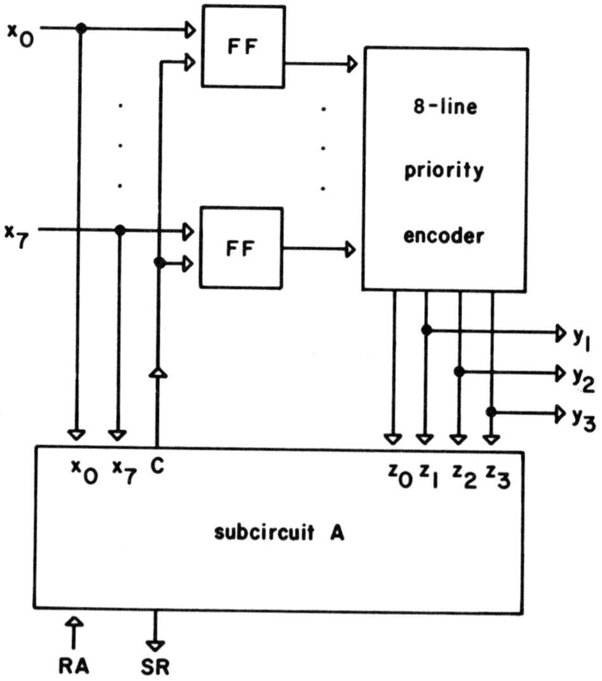

Figure 3.19 Decomposition of request controller

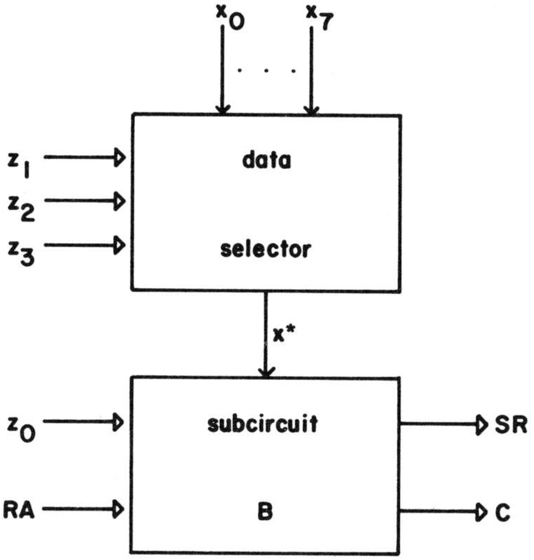

Figure 3.20 Subcircuit *A* decomposed

the device and the controller). The controller is now ready to take care of the next awaiting service request.

In this particular design example, n will be taken to be equal to 8 so that we can have a maximum of eight peripheral devices to be connected and thus m will be equal to 3. As a result, an 8-line priority encoder (Problem 2.21) can be used to produce the address lines y_1, y_2 and y_3 from the input lines $x_0, x_1, \ldots, x_7$.

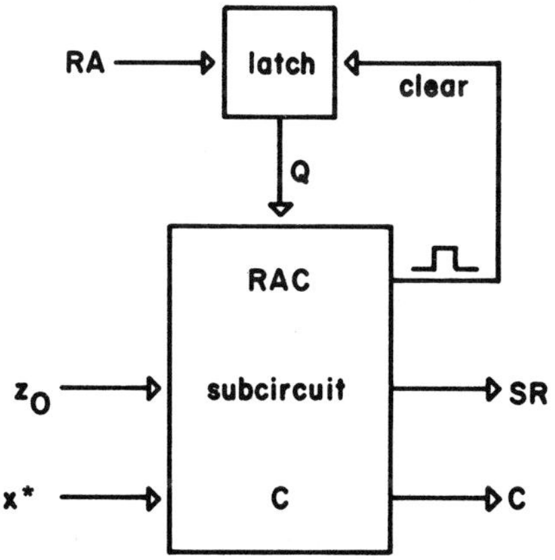

Figure 3.21 Subcircuit *B* decomposed

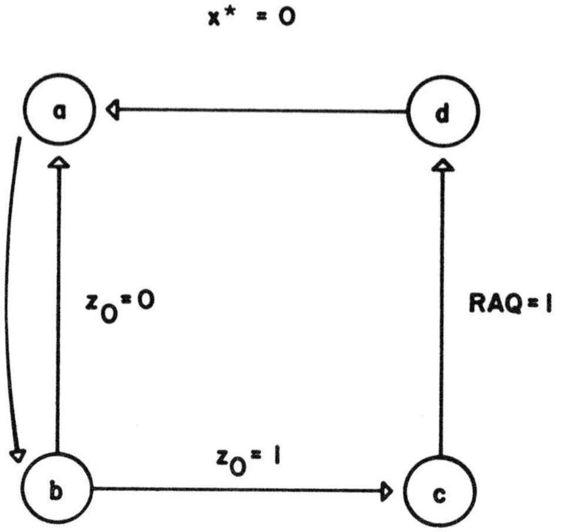

Figure 3.22 State diagram for subcircuit C

Incorporation of the 8-line priority encoder yields in block diagram form the circuit shown in Fig. 3.19 where subcircuit A has yet to be designed. The flip-flops are used to store the values of the x_i's. The clock signal C produced by subcircuit A is used to sample the values of the x_i's at appropriate moments to be described in more detail when the subcircuit A is represented. Informally, the function of subcircuit A is to monitor a selected request until the request disappears. Subcircuit A has inputs z_0, z_1, z_2, z_3, x_0, ..., x_7 and RA and outputs C and SR.

Using a data selector or multiplexer (Problem 2.22) subcircuit A can be decomposed as shown in Fig. 3.20. The data selector produces an output x^* which is equal to x_i, the signal at the request line for the device with address i whose request the controller has been servicing. Subcircuit B is the control structure with inputs x^*, z_0, RA and outputs SR and C. We now start to design subcircuit B as an asynchronous sequential circuit.

We first overcome the difficulty that the signal RA is an input pulse

State	Condition
a	$c \leftarrow$ ⎍ ; $RAC \leftarrow$ ⎍ ; $SR \leftarrow 0$
b	no output signal
c	$SR \leftarrow 1$
d	no output signal

Table 3.22 Interpretation of states in Fig. 3.22

which occurs independently of the internal clock of subcircuit B by the introduction of a latch (flag) as shown in Fig. 3.21. Subcircuit C is now represented in the form of the state diagram shown in Fig. 3.22 where the states are used to represent conditions as shown in Table 3.22. Note that a slight variation in convention has been used to indicate the various output signals produced in the drawing of the state diagram and this should not produce any confusion.

Transition from one state to another is assumed to take one clock period. If the transition actually requires a delay of more than one clock period, it is necessary to introduce the appropriate number of dummy states between the transition from that state to the other state in order to obtain the desired amount of delay.

The implementation of subcircuit C is now straight forward and will be left as an exercise to the reader (Problem 3.11).

Let us look at a third example. In this example, a simple combinational lock is to be designed. A sequence of three binary-coded-decimal (BCD) digits acting as the lock's combination is assumed to be stored in a memory unit in the circuit. A BCD code is a 4-bit binary representation of a decimal digit 0 to 9. Three binary-coded-decimal digits acting as the key is then taken as a sequence of inputs to the circuit. If the two sequences match, then an output of 1 is produced by the circuit which can then be used to activate a device to open the lock.

A block diagram of the combinational lock is shown in Fig. 3.23. The input digit in *BCD* code comes in through the lines X_0, X_1, X_2 and X_3. The input R is used for resetting so that it is high when a new cycle or sequence of input digits is to be made. Line P is used to convey a pulse which indicates the presence of an input digit on X_0, X_1, X_2, and X_3. Z is the output terminal. After the restart terminal has been reset, the first *BCD* digit present at X_0, X_1, X_2, and X_3 is assumed to be the first digit of the key to be tried. Then the second and third digits are tried in order.

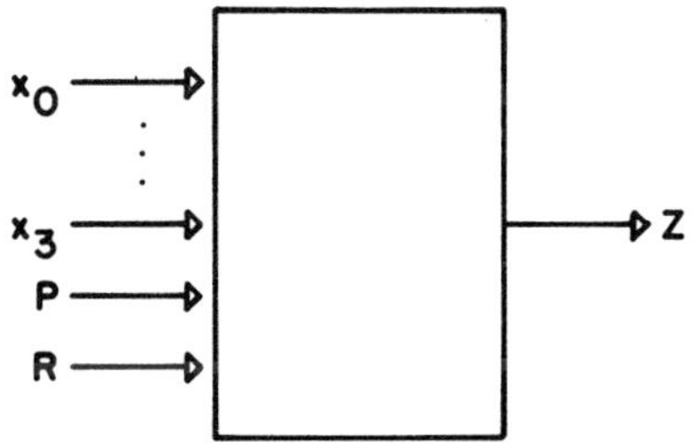

Figure 3.23 Combinational lock

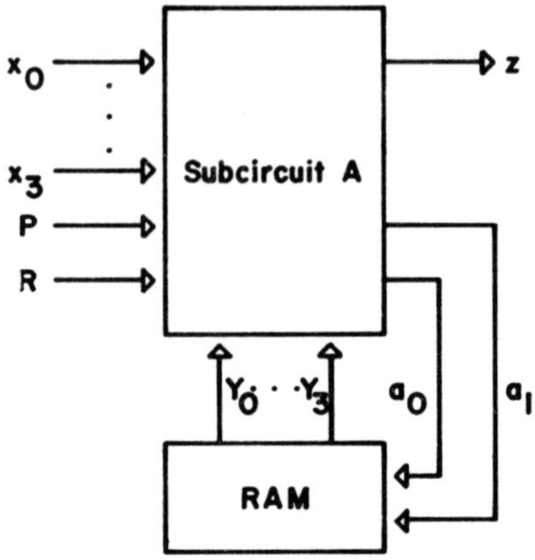

Figure 3.24 RAM unit incorporated

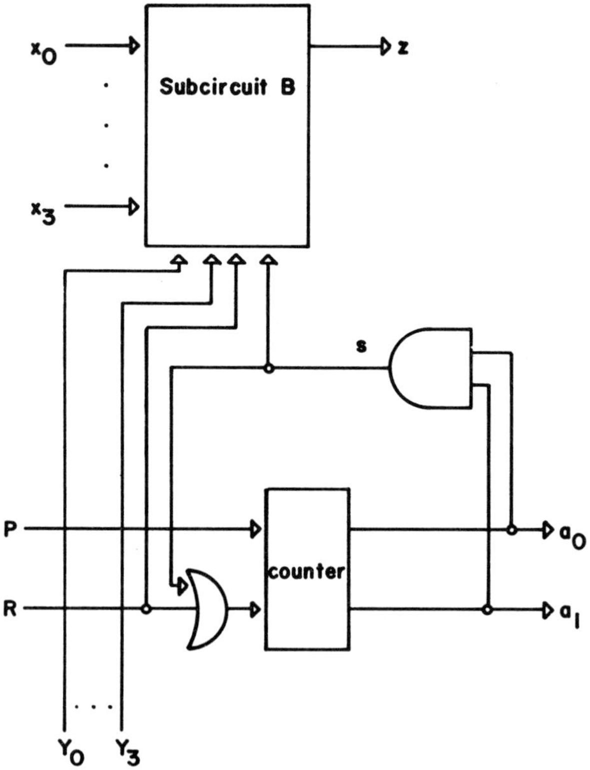

Figure 3.25 Subcircuit *A* decomposed

If the three digits match the three stored digits representing the lock combination, a 1 will be produced as output at terminal Z, otherwise a 0 will be produced.

To store the lock combination, it is convenient to use a small RAM which is widely available as an IC circuit. Using a RAM for this purpose enables the lock combination to be changed. We use a RAM unit which can store three *BCD* digits. A RAM unit which has four words of 4-bit length is adequate. Assuming the RAM has the necessary addressing circuitry, we have in block diagram form the circuit shown in Fig. 3.24. The signals a_0 and a_1 are lines for addressing the RAM unit and the addressed word storing the corresponding digit is read out on lines Y_0, Y_1, Y_2 and Y_3.

To produce an address for reading out the correct combination digit, a counter is used. Incorporation of a counter decomposes subcircuit *A* as shown in Fig. 3.25. Note that the outputs from the counter are also ANDed together to form a signal *s* which will be used to gate the output of the entire circuit into terminal 2 at the end of the cycle, i.e., when all three input digits have been compared, and also to clear the counter to prepare it for the next cycle. The counter counts the number of pulses coming in from *P* and can be manually reset from *R*.

Incorporation of a flip-flop, denoted *FF*, and two AND gates decomposes subcircuit *B* into that shown in Fig. 3.26. It is not difficult to see how subcircuit *C* is to operate: the signals on X_0, X_1, X_2, and X_3 are compared with those on Y_0, Y_1, Y_2, Y_3 respectively and *FF* is set if and only if, $X_0 = Y_0$, $X_1 = Y_1$, $X_2 = Y_2$ and $X_3 = Y_3$ and its initial value is 1. Note that the restart signal *R* is used to set *FF* to 1 at the beginning of each cycle. The realization of subcircuit *C* is trivial. Thus, each time

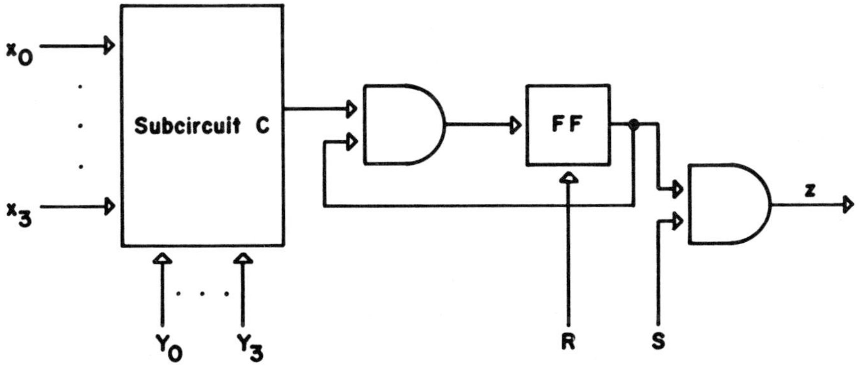

Figure 3.26 Subcircuit *B* decomposed

a *BCD* digit is made available to X_0, X_1, X_2, and X_3 accompanied by a pulse on *P*, the corresponding combination digit is read out from RAM onto Y_0, Y_1, Y_2, and Y_3 and compared with the input *BCD* digit. *FF* is appropriately set to 1 or 0 and an output is produced at *Z* only when all three digits have been compared.

3.7 SHIFT REGISTERS AND COUNTERS

A register is a device containing a set of flip-flops used for storing a digital number and for performing operations on that stored number. Registers take on a number of forms: for example, the decade-pulse counter discussed in Section 3.3 is a specific kind of register. Two other kinds of registers, the shift register and the counter, will be studied in this section. These special purpose registers are widely used in many digital circuits and are widely available as low-cost ICs.

A 4-bit shift register using *J-K* flip-flop is shown in Fig. 3.27. On the application of a pulse at the *T* terminals, the content of flip-flop *A* is shifted into flip-flop *B*, that in *B* into *C*, that in *C* into *D*. Thus the value of bit b_3 will appear as bit b_0 after three shift pulses. One obvious application of this shift register is the implementation of delays in a circuit. A value loaded into *A* can be made to appear as the content in *D* after a specified amount of time by controlling the speed of the clock used to generate the shift pulses. This is called a unidirectional shift register. A bidirectional shift register can shift in either of two directions depending on the value of a control signal.

Connecting the output of flip-flop *D* to the input terminals of flip-flop *A* results in a ring counter (or rotate register) as shown in Fig. 3.28. The content of each of the flip-flops will be restored after the application of

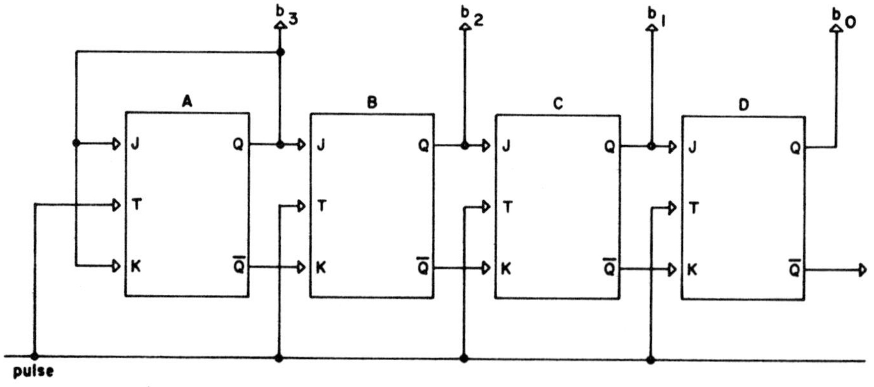

Figure 3.27 A 4-bit shift register

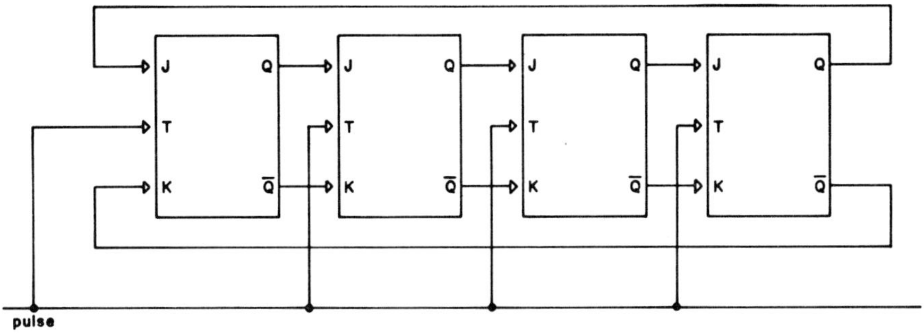

Figure 3.28 A 4-bit ring counter

four shift pulses. A ring counter built this way can be used to generate a cyclic sequence of bit patterns. For example if the register originally contains 1001, then the ring counter generates the sequence 1001, 1100, 0110, and 0011 cyclically. Cyclic sequences are very useful in designing digital circuits where fixed sequences of events are to be repeated.

A more general ring counter can be obtained through the use of a shift register and some additional logic as shown in the block diagram of Fig. 3.29. The content of each of the flip-flops in the shift register is used as input to the combinational-control circuit to generate an input I to the first flip-flop of the shift register. In this way, cyclic sequences of length less than or equal to 2^n, where n is the number of flip-flops in the shift register, can be implemented. As an example, to implement a ring counter generating the following cyclic sequence:

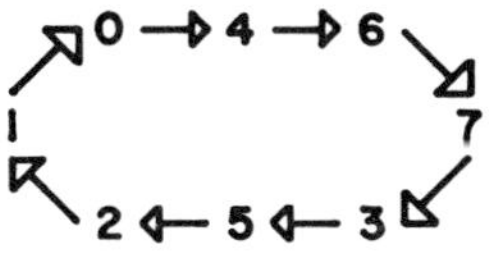

a shift register with $n = 3$ is used. A truth table for the combinational-

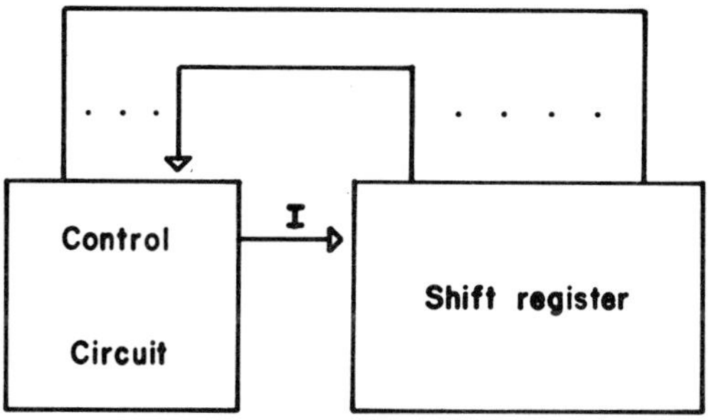

Figure 3.29 General configuration of a ring counter

b_2	b_1	b_0	I
0	0	0	1
0	0	1	0
0	1	0	0
0	1	1	1
1	0	0	1
1	0	1	0
1	1	0	1
1	1	1	0

Table 3.23 Truth table for control circuit of ring counter

control circuit is shown in Table 3.23. The contents of the three flip-flops in the shift register are denoted as b_2, b_1, b_0 and I is the input to the first flip-flop. For example, when state 6 is entered, the content of the shift register is 110, The next state is to be 7. To obtain a content of 111 for the register after shifting 110, a 1 has to be applied at the terminal I accounting for the entry of 1 for the combination of $b_2b_1b_0 = 110$ in the truth table. The rest of the table can similarly be deduced. The realization of the control circuit is straightforward.

A counter that is capable of counting from 0 to $N - 1$ repeatedly, where N is an integer, is known as a modulo-N counter. For example, the decade counter described in Section 3.3 is a modulo-10 counter. The design procedure for the decade counter can be used to obtain a modulo-N counter for arbitrary N. However, modulo-K where K is a power of 2 can be implemented much more easily, for example, using trailing-edge triggered flip-flops. A flip-flop is said to be trailing-edge triggered if the state transition (from 1 to 0 and vice versa) of the flip-flop occurs whenever the

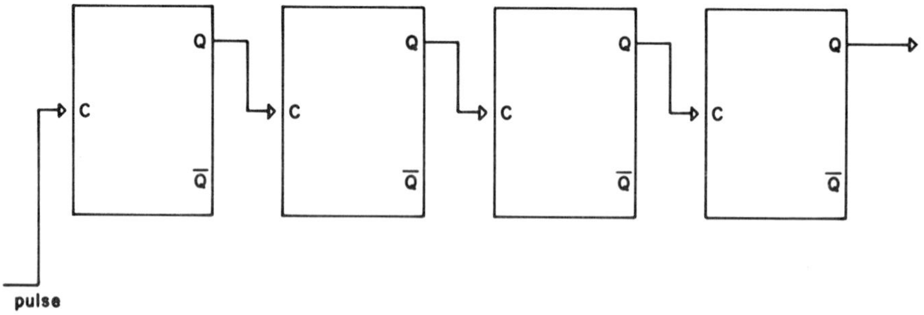

Figure 3.30 A modulo-16 counter

clock input changes from high to low. A modulo-16 counter implemented using this type of flip-flop is shown in Fig. 3.30.

PROBLEMS

3.1 For the machine defined by the following diagram:

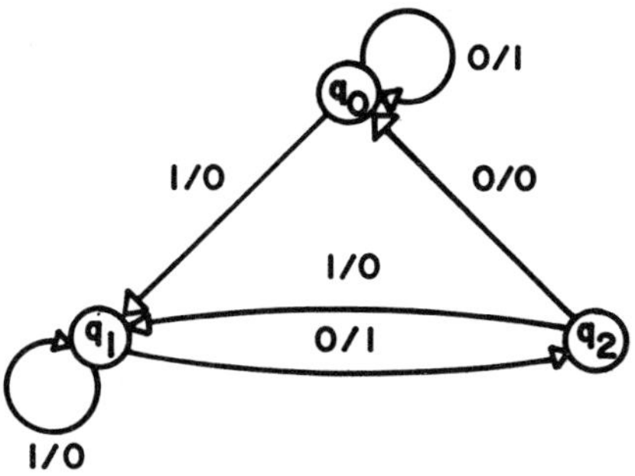

find

(a) $\delta(110011,q_0)$, $\delta(00110010,q_1)$
(b) $\omega(110011,q_0)$, $\omega(00110010,q_1)$

3.2 Find the optimal machine for the machine represented by the following state table:

Present state	Next state 0	Next state 1	Output 0	Output 1
q_0	q_1	q_2	0	0
q_1	q_3	q_4	0	0
q_2	q_5	q_6	0	0
q_3	q_7	q_8	0	0
q_4	q_0	q_0	1	0
q_5	q_0	q_0	0	1
q_6	q_0	q_0	1	0
q_7	q_0	q_0	0	1
q_8	q_0	q_0	1	0

3.3 Draw a state table for the machine in Problem 3.1. Is the machine an optimal machine?

3.4 For the following incompletely specified sequential machine, identify the maximal compatibles and then the prime compatibles:

Present state	Next state 00	01	11	10	Output 00	01	11	10
q_0	q_1	—	q_2	q_1	0	—	—	0
q_1	q_2	q_4	q_1	—	0	1	0	1
q_2	q_2	q_3	—	q_4	0	1	—	0
q_3	—	q_0	q_1	—	—	1	—	—
q_4	—	—	q_0	—	—	—	1	—

Construct the optimal machine.

3.5 Construct the state diagram for a synchronous sequential circuit which can be used to detect errors in 2-out-of-5 coded messages. The machine examines the messages serially and produces an output of 1 whenever an illegal message (the message is legal if and only if exactly two of its bits are 1s) of five binary digits is detected.

3.6 The following state table is to be realized using two *T* flip-flops:

Present state	Next state 0	1	Output 0	1
A	B	B	0	0
B	C	A	0	1
C	B	D	0	0
D	C	B	0	1

Choose an appropriate secondary assignment and give the corresponding excitation and output switching expressions.

3.7 For the following next-state table an asynchronous sequential circuit is to be built (Note that there are three inputs I_0, I_1, and I_2):

Present state	Next state I_0	I_1	I_2
q_0	q_0	q_1	q_2
q_1	q_0	q_1	q_3
q_2	q_0	q_2	q_3
q_3	q_0	q_2	q_3

Using two state variables, propose a secondary assignment which can result in a critical race-free circuit.

3.8 A given machine is represented by the following state diagram:

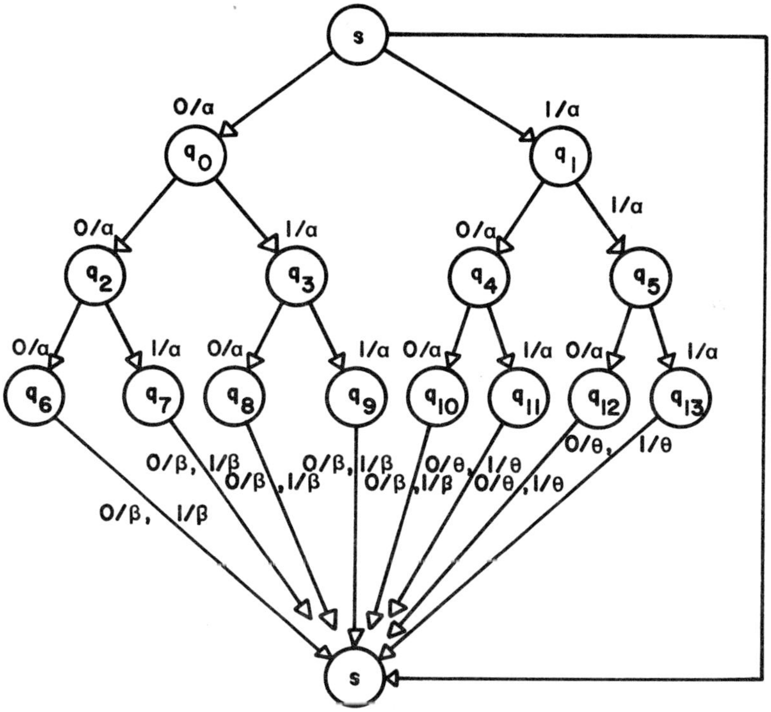

1. Draw the corresponding state table for the given machine.
2. Find an equivalent optimal machine (give the state table).
3. Briefly explain what type of computation the machine may perform.

3.9 Following the same procedure as used in the design of the synchronous detector for the pattern 101 discussed in this chapter, design a synchronous detector for the pattern 011.

3.10 Implement subcircuit C in Fig. 3.22 using JK flip-flops.

3.11 In each of the following cases, implement a ring counter having the given cyclic sequence using a 3-bit shift register and a multiplexer as the control circuit:

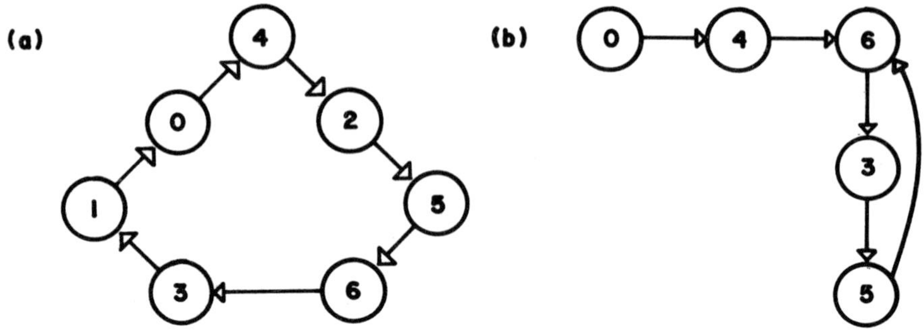

REFERENCES

1. Blakeslee, T. R. *Digital Design with Standard MSI and LSI*. New York: John Wiley & Sons, 1975.
2. Booth, T. L. *Sequential Machines and Automata Theory*. New York: John Wiley & Sons, 1967.
3. Brzozowski, J. A., and Yoeli, M. *Digital Networks*. Englewood Cliffs, (N.J.): Prentice-Hall, 1976.
4. Clare, C. R. *Designing Logic Systems Using State Machines*. New York: McGraw-Hill, 1973.
5. Hill, F. J., and Peterson, G. R. *Introduction to Switching Theory and Logical Design*. New York: John Wiley & Sons, 1968.
6. Kohavi, Z. *Switching and Finite Automata Theory*. New York: McGraw-Hill, 1970.
7. McCluskey, E. J. *Introduction to the Theory of Switching Circuits*. New York: McGraw-Hill, 1965.
8. Torng, H. C. *Switching Circuit Theory and Logical Design*. Reading, (MA): Addison-Wesley, 1972.
9. Wood, P. E., Jr. *Switching Theory*. New York: McGraw-Hill, 1968.

Chapter 4

DESIGN OF DIGITAL COMPUTER COMPONENTS

4.1 INTRODUCTION AND OVERVIEW OF DIGITAL PROCESSOR

In Chapter 2, we discussed how arithmetic or logical problems can be solved by implementing a suitable combination of switching circuits. This technique can be used to solve very simple or complex problems. A non-trivial problem or operation can usually be broken down into a sequence of simple steps. These steps may then be organized to perform in some specified sequence to solve the overall problem. Different problems would require different sequences of operations to solve them. Thus, we may provide hardware to implement a basic set of arithmetic or logical operations, and a mechanism for establishing the sequence in which these hardware components will operate. We developed the theory and designed specific sequential circuits in Chapter 3. In that chapter, machines were designed to solve a problem by following a sequence of steps. In each instance, the machine decided what step to follow next based on current input(s) and the result of previous operations. The overall solution depended on the starting state and the external input sequence. If the machine is set to the same starting state, then identical input sequences will produce identical output results.

Another important point to consider here is the choice of building blocks for logic circuits. In Chapters 2 and 3, we discussed circuit implementations using basic gates and flip-flops. Any circuit can be implemented using these building blocks or components. A reasonably complex circuit will require a large number of these components which must be appropriately connected to perform the desired operation. The space required by these components on a circuit board may be quite substantial and the physical layout may become quite involved for establishing proper interconnections; the power dissipation and loading requirements of all the components must be considered and satisfied.

In the late 1960s, many manufacturers of integrated circuits started production of monolithic circuits (on a single silicon chip) with several gates or flip-flops encapsulated in a single package with appropriate input-output contacts (leads). This packaging density (number of logic gates or flip-flops serving as storage elements) has rapidly increased. With the increase in this packaging density made possible by improvements in technology, entire logic circuits were implemented on a single chip thus reducing the cost and power dissipation per component. An integrated circuit chip with 100 components may cost no more and consume no more power than a chip with a single gate or flip-flop. This has led to redundant design techniques, i.e., use of chips which provide higher level of circuit complexity than the minimum requirements of a given design. With appropriate signals at the input and control leads of the IC chip, the circuit can be made to perform the desired operation. A typical example is the use of data selector ICs to implement switching functions (see Section 2.9).

Therefore, while the techniques presented in Chapters 2 and 3 still form the basis, and are directly applicable in many instances, most arithmetic and logic circuits of a digital computer can now be provided by choosing a small number of ICs or LSI chips of a much higher level of complexity, permitting simple physical layout and connections on the circuit board.

Consider, for example, the problem of adding two 8-bit binary numbers. We can construct a combinational logic circuit which receives a total of 16 inputs (8 bits for each number) and produces the sum which appears on eight or nine output lines (two 8-bit numbers produce a 9-bit sum for some input values) (Fig. 4.1). On the other hand, we could bring the input lines to two separate holding areas (registers), use a 1-bit adder circuit to perform addition on the consecutive bits of the input data, and produce corresponding bits of the result sequentially (Fig 4.2). In this case the single addition operation has been changed into a sequence of 8-addition operations. The following comments provide additional insight:

1. The combinational circuit requires a single-enabling (clock, gating)

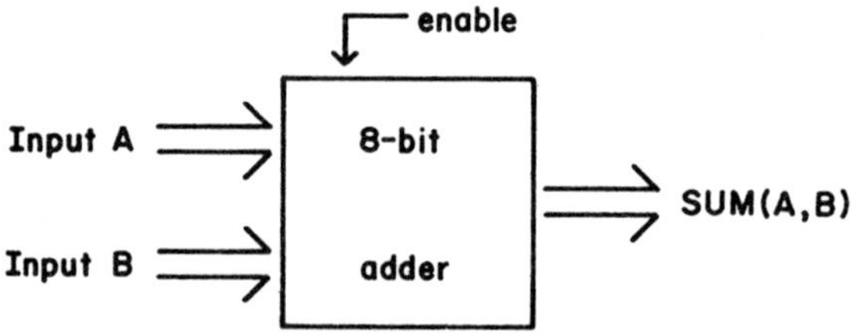

Figure 4.1 An 8-bit combinational addition circuit

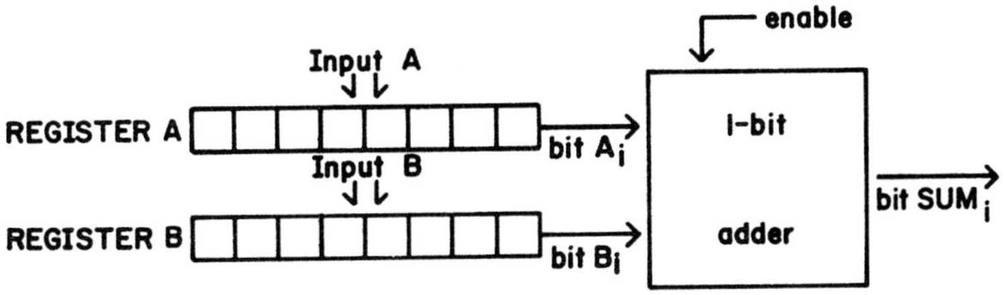

Figure 4.2 An 8-bit sequential bit-by-bit addition circuit

pulse for the adder, and this can be done soon after the signals on input lines *A* and *B* (Fig. 4.1) are stable, or at any time during which these signals retain desired inputs. The result on the output lines appears almost instantaneously after the enable signal. All 8-bits are added in parallel, simultaneously.

2. The corresponding sequential circuit requires a sequence of 8 enabling pulses with suitable intervals to perform the complete addition serially, bit by bit.
3. While the combinational circuit requires very little or no control logic to perform addition, the sequential circuit requires considerable control logic and hardware. Additional hardware consists of registers *A* and *B* capable of shifting the bits required for addition, and a register for the output (to retain all bits of the result). The control sequence consists of the following basic steps:
   a. Load *A*; Load *B* (Note that *A* and *B* may be loaded simultaneously with a single control pulse.)
   b. Generate a sequence of 8 pulses to enable the 8 addition operations. Shift the result of addition into output register for each bit.
4. The sequential circuit will take more time to perform addition. However, the addition circuit used in this case is much simpler and less expensive.

If we were designing a machine just to add two numbers, then the combinational approach would be adequate. However, in a digital computer we usually provide many basic operations, such as addition, subtraction, comparison, shifting, etc. In this case registers would be shared by all of these circuits and overall cost is reduced. Also, many new operations may be performed by using the circuits already available and simply modifying the control sequence.

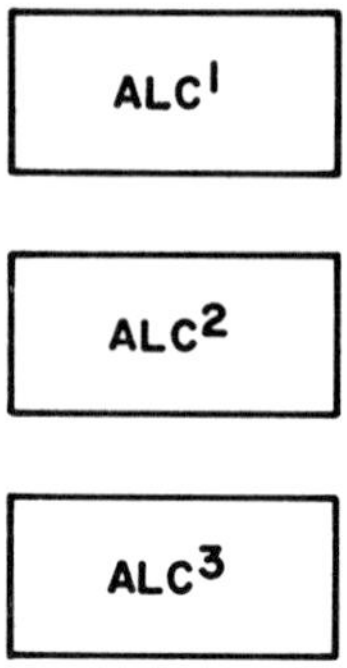

Figure 4.3 Representation of arithmetic or logic circuit blocks

For example, to perform subtraction, the following sequence can use the adder circuit already available by 2's complement arithmetic:

(i) Load A; Load B

(ii) Complement the contents of B and load into B

(iii) Perform bit-by-bit addition

Similarly, multiplication can be performed by repeated use of the addition circuit. Of course, we could retain the 8-bit adder and provide it with appropriate control signals to do multiplication. Thus the difference between combinational and sequential circuit is not simply that one operates on input-signal bits simultaneously (in parallel) and the other operates on signal bits in serial fashion. A sequential circuit has a control sequence which selectively enables other circuits to perform the desired operation. These other circuits may themselves be combinational or sequential in nature.

Let us now return to our discussion of implementing arithmetic and logical operations for a digital computer processing unit. Let these arithmetic or logic circuits be denoted by ALC^1, ALC^2, ALC^3, etc., and shown as logic blocks in Fig. 4.3.

If the operation is unary in nature, i.e., requires only one operand, then we may assume that this single operand is contained in a register which will be called the *accumulator*.

In order to perform the desired operation we establish a data path from AC to an appropriate circuit block ALC^i as shown in Fig. 4.5. ALC^i is continuously performing the operation on its inputs. We must now decide to where the output of ALC^i should go. Generally, results of operations

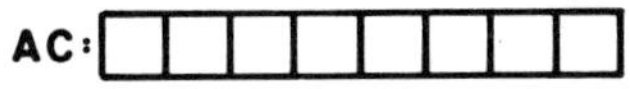

Figure 4.4 Accumulator register for holding one of the operands in arithmetic and logical operations

are routed to the accumulator as shown in Fig. 4.6. Now we must consider the problem of timing or sequencing. We would like the data to be transferred from AC to ALC^i when an appropriate signal to this effect is issued and we want to allow ALC^i enough time to complete its operation before it transfers the result of the operation to the accumulator. Thus, we will generate a pulse at time t, and use it to clock (gate) the contents of AC to ALC^i; route this pulse through an appropriate delay element, and use the delayed pulse to clock the output of ALC^i into AC. This is shown schematically in Fig. 4.7. The above sequence of operations can be represented as follows:

$$\begin{aligned} &1: \quad ALC^i \leftarrow AC \\ &2: \quad AC \leftarrow ALC^i \end{aligned} \tag{4.1}$$

In this representation we assume that if a transfer operation for a step n takes place at time t, then the transfer corresponding to step $n + 1$ will take place (clocked) at time $t + 1$ unless specified otherwise.

As an alternative to the configuration shown in Fig. 4.7 we may leave a direct (unclocked) connection from AC to ALC^i, so that ALC^i would be continuously operating on whatever data AC contains at any given time. However, we control the flow of the result into AC, resulting in the configuration shown in Fig. 4.8. This sequence of operations can be represented as

$$\begin{aligned} &1: \quad \text{Load } AC \text{ with source data} \\ &2: \quad AC \leftarrow ALC^i \end{aligned} \tag{4.2}$$

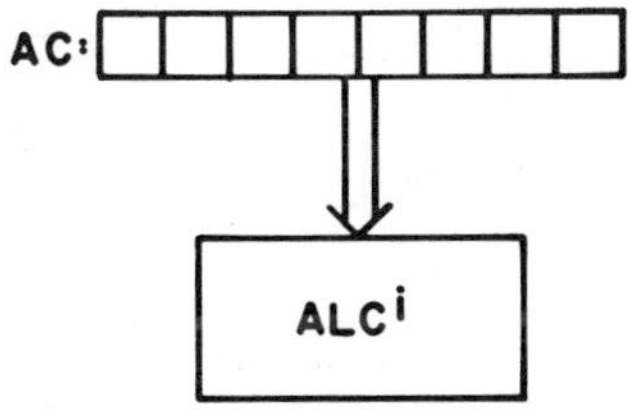

Figure 4.5 Accumulator register as source data to arithmetic and logic circuit

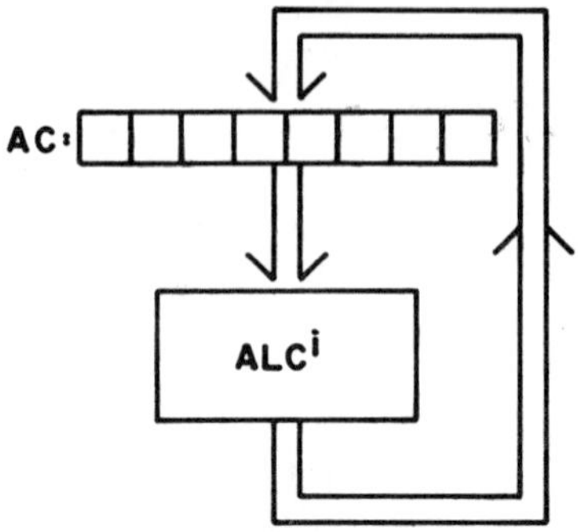

Figure 4.6 Accumulator register as source to and destination from arithmetic and logic circuits

Thus the accumulator is either loaded with the source data or the output of ALC^i depending on the corresponding clocking operation. In many operations two operands are required. Let the second operand be contained in another register, called B. Also, a 1-bit register, i.e. a single flip-flop, is required on the left of the accumulator, AC, to accommodate an extra bit in arithmetic operations. Let ALC^i be an adder circuit, for example. The resulting organization of components and clocking is shown in Fig. 4.9. Using this configuration a two-operand operation can be represented as follows:

1: Load AC with first operand

2: Load B with second operand

3: $c, AC \leftarrow ALC^i, 4\Delta$ (4.3)

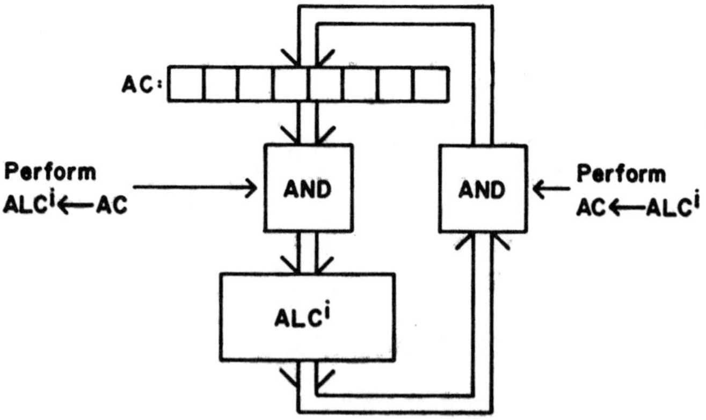

Figure 4.7 Clocked transfers of data

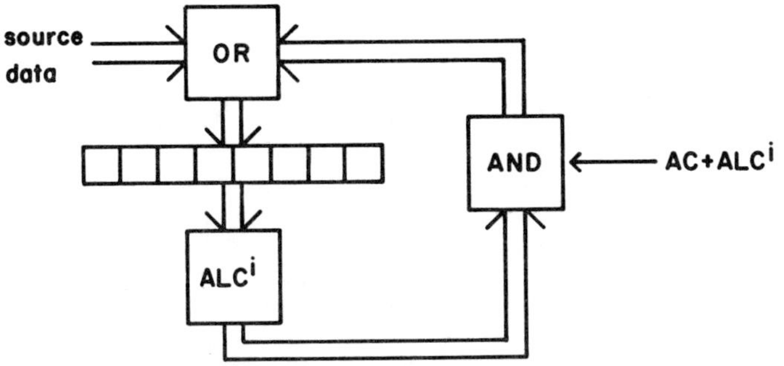

Figure 4.8 A variation on configuration shown in Fig. 4.7

In step 3 the clocking of the adder output into c,AC is delayed by four units of time since we assume that the adder circuit will take this much time to perform the operation before the correct result appears on the output.

We have shown two examples of arithmetic and logic circuits and two registers for holding data required in using these circuits. In general, the arithmetic and logic unit (ALU) of a computer contains many such circuits for different operations and a number of additional registers required for holding other relevant data.

The input or source data used in performing arithmetic or logic operations can come from a number of sources. For example:

1. A set of binary (ON, OFF, or 1,0) switches. By opening and closing (ON, OFF) operations of these switches we may provide bit patterns consisting of 1s and 0s. Switch contacts bounce from a few to a few tens of msec before settling down. Therefore, rather than using the outputs of these switches directly we may connect the output of

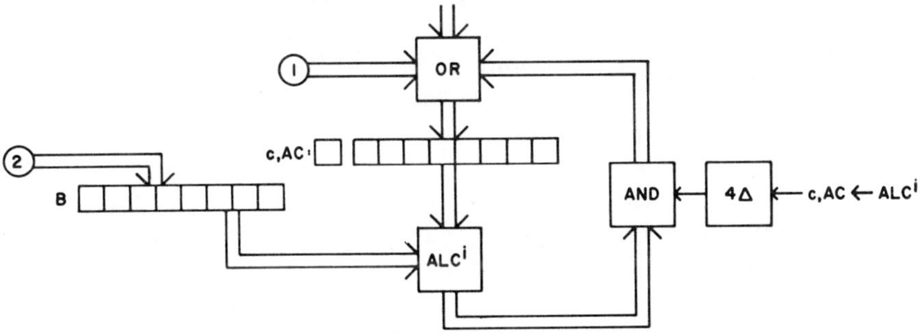

Figure 4.9 Organization of components for operations requiring two operands

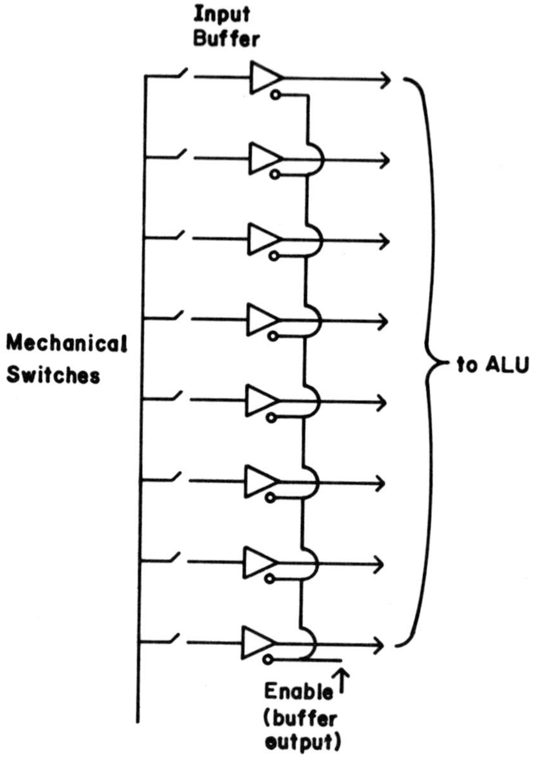

Figure 4.10 Mechanical switches connected to an input buffer. The input buffer levels as set by switches are passed to ALU when an enable signal is issued.

switches to buffers by clocking whenever we want to pass the values set by the switches to the other circuits. Buffers are storage elements for receiving and holding an input bit pattern from some source. This bit pattern is subsequently picked up by the intended destination. Buffering provides compatibility of signal levels and also eliminates the need for synchronization between the source device and the destination device. A given buffer register consisting of storage elements for one or more bits can be used by any one of the selected sources to store data at one time and read by any one of the selected destinations at a later time. This configuration is shown in Fig. 4.10. The input buffer shown in Fig. 4.10 consists of two commercially available ICs (74125, Quadruple bus buffer gates with 3-state outputs). Gates with tri-state outputs are important whenever several devices must be connected to a common bus (a set of lines for routing of data). These gates, when enabled, cause the appearance of their high or low signal levels on the bus. The third state (when gates are not enabled) known as the "floating" state avoids loading of the bus as if these gates were temporarily disconnected from the bus.

2. The input or source data may also be received from a number of other devices such as teletype, keyboard, punched card, punched paper tape, etc. However, in these cases the interconnection of these devices to the ALU is more complex and the signaling is more involved. We will defer the detailed discussion of these devices to Chapter 8.

3. The input or source data may have been received earlier from any of the devices mentioned above, and stored in a memory device. There are many reasons for using this scheme. For example, it may not be possible to receive source data from an input device precisely at the time that it is needed for arithmetic and logic operations. Also, input devices involving mechanical operations are very slow compared to the speed at which electronic arithmetic and logic circuits operate. Thus if a sequence of arithmetic and logic operations has to be performed, the electronic circuits would have to wait before each operation to receive data from mechanical input devices. On the other hand, it is possible to design memory devices which will respond very rapidly to requests for data items stored in them. Now the information from the input devices can be received when it appears and passed to selected locations in a memory or storage device. Later, when this information is needed, an appropriate memory location (address) is provided. On the application of an appropriate signal, the memory device retrieves information from this address and outputs it to a memory output buffer. The information from this buffer can be routed to the ALU.

The ALU can, therefore, receive information from one of the many input devices, including the memory or storage unit. Rather than having several independent paths going into the ALU from input data sources, inputs will usually be brought to a common interconnection and from this common interconnection one data path is established to the ALU. This is known as a common data bus. Similarly, as a result of operations in the ALU, information may be passed to output devices. Again a common data bus is used. Usually a bidirectional bus that can transmit information in both directions is established as shown in Fig. 4.11. This bidirectional bus provides a set of common lines for routing of data from the ALU to

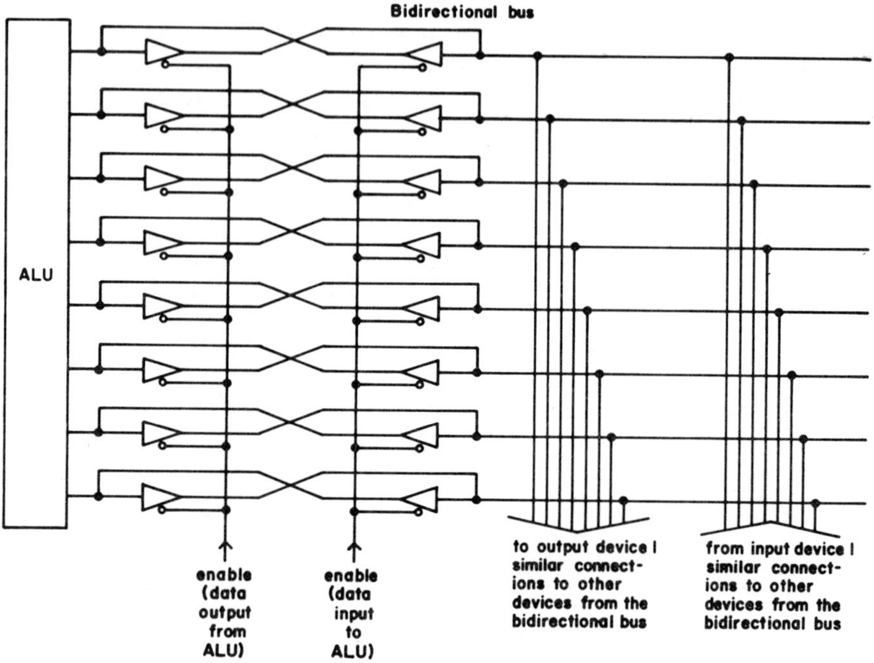

Figure 4.11 ALU vs. input-output data flow and signaling

other circuits when the data output line is enabled, and routing of data from other circuits to the ALU when the data input line is enabled. Thus, the direction of data flow can be established by a control unit or device which enables the input or output line, whichever is desired at a given time.

So far we have considered single operations in the ALU and what is required to execute these single operations. Generally, solution of a problem would require a suitable sequence of these single operations. We could perform this sequence manually. However, in this case the speed of solution would correspond to manual speed, and essentially we would be using our ALU as one of the commercially available simple calculators. For long sequences this procedure becomes very tedious, particularly if a given sequence of operations is to be used for different sets of data values. In this case we would be reinserting every step of a given sequence again and again and providing the data values. In many cases, the problem may require some but not all data values to be changed, but we will have to reinsert all of them again unless some space is provided to store the data values and modify them selectively. Similarly, in our sequence of operations, we may want to modify some steps, or delete a few steps, or add a few new steps to solve a new problem or the same problem differently. In all these cases we will have to go through manually reinserting the complete sequence step by step. These situations clearly demonstrate the need for a storage device which can be used to store a sequence of operations for the problem just as we store data. We can use the same storage device to store operations (instructions for operations) as well as data, and employ pointers to or addresses of instructions to be executed next (operation to be performed next) as well as data value(s) to be retrieved for this operation. It is possible for us to establish two pointers, one for the instruction location and one for data location. We may maintain these two pointers in the ALU. However, if the same storage device is used for storing instructions and data, then the storage device will have an address register associated with it which will either point to an instruction being retrieved (accessed) or data being retrieved (accessed). The storage device can bring the information retrieved into a buffer register. The contents of this buffer register can then be interpreted as an instruction or data depending on the original request.

The preceding paragraph implies that there is a need to augment the ALU with two additional features: sequencing through a set of instructions, and interpretation and translation of a retrieved instruction into a sequence of operations of the arithmetic and logic circuits of the ALU. The ALU thus augmented is known as a *processing unit* or *central processing unit* because the overall computer configuration may include additional specialized processors (processing units) for some common but complex operations related to input-output and other activities.

4.1.1 Processor Instruction Sequencing

Let us assume, for a moment, that instructions as well as data have been previously stored in the storage device called *MEMORY,M*. Associated with *M* are registers MA and MB as mentioned earlier. A basic set of processor registers is also defined:

MA: a memory address register which contains the address of the location to be accessed. The size of this register (number of bits) depends on the number of memory locations, e.g., 8 bits will provide address locations 0 to 255, 10 bits will provide address locations 0–1023 (generally known as memory of size 1K), 12 bits will provide all addresses for 4K (4096) memory, etc.

MB: a memory buffer register which will contain the information at the location specified by MA. The number of bits in this register corresponds to the number of bits contained in each memory location (memory word).

PC: a program counter in the processor which contains the address of the next instruction to be executed. Its size corresponds to MA.

IR: an instruction register in the processor which contains the bit pattern corresponding to the instruction currently under execution. Its size corresponds to one or more memory words.

AC: a register in the processor which contains one of the source operands as defined previously; the second operand may be picked directly from MB (shown as *B* before). The size of AC has already been indicated.

c: a one bit register as defined previously, in sequence (4.3).

The hardware configuration is shown in Fig. 4.12. A program consisting of a set of instructions will be executed in its entirety (from first instruction to last) by the sequencing shown in Fig. 4.13 to be performed by a sequencer (controller) contained in the processor.

Assuming for the moment that every instruction is a valid instruction (i.e., it can be interpreted and executed without any difficulty which would require interruption of the program), the flowchart shown in Fig. 4.13 needs a few more details. The instruction currently being executed may specify a branch to an instruction other than the next one in sequence, and in this case, PC, should be appropriately modified. Also, this instruc-

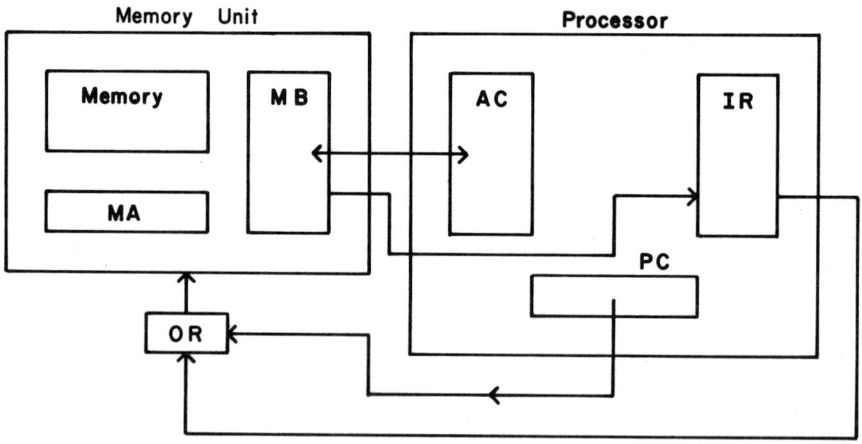

Figure 4.12 Basic hardware configuration showing processor and memory unit

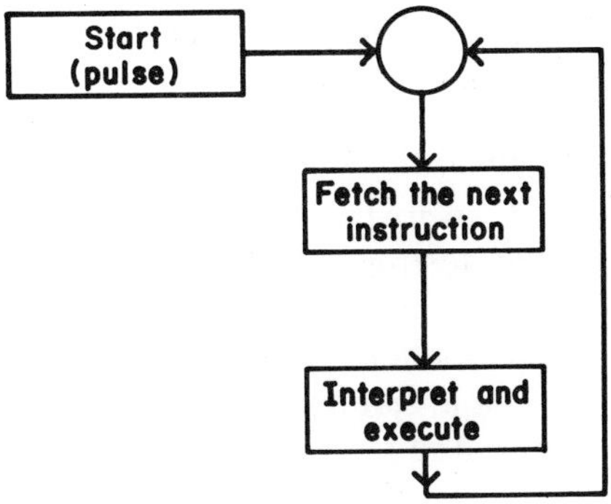

Figure 4.13 Basic instruction sequencing

tion may be the last instruction indicating a halt operation rather than going to next instruction. Both of these possibilities would be discovered during interpretation and execution. Thus the flowchart can be modified as shown in Fig. 4.14.

The flowchart shown in Fig. 4.14 assumes that PC is loaded appropriately (address of the instruction to be executed) before the *start* button is pressed. How the branch address is determined will be explained when

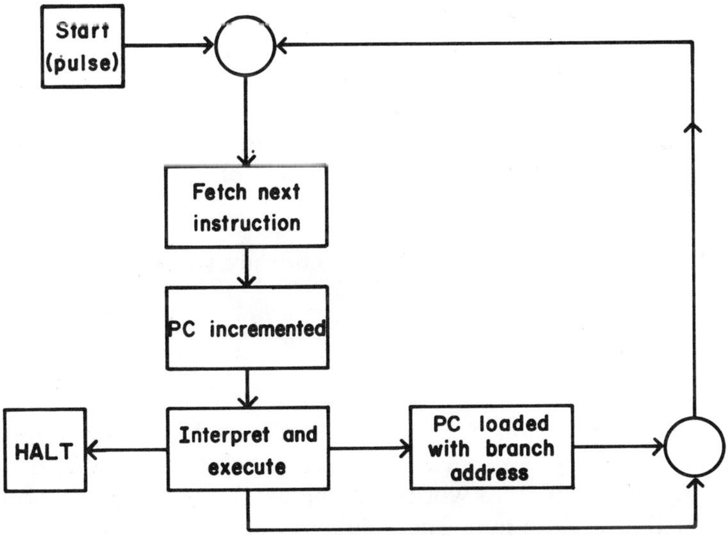

Figure 4.14 Basic instruction sequencing modified to include branching and halt

we discuss instruction interpretation. At that time we will also describe an appropriate sequence algorithm.

4.1.2 Instruction Decoding and Execution

As mentioned earlier, an instruction is used to activate one or more ALU circuits so that the desired arithmetic or logical operations are executed. Different instructions are represented by different binary bit patterns and are written and stored in memory. If each memory word consists of 8-bits (as we have been assuming so far) then one full word or part of it may be used to express the operation code. Using all 8-bits, a total of 256 operation codes are possible. However, for each operation we need to specify one or more operands (data) implicitly or explicitly. For example, consider an instruction to load a data item stored in memory. If the accumulator AC will always serve as a register in which data value is loaded, then this is implicit and need not be identified in the operation code. However, the address in memory from which the data value is to be loaded has to be stated explicitly at least once so that an arbitrary data value may be accessed from memory. Of course, for some instructions it may be possible for us to put the data value immediately after the instruction in memory, and in this case a different operation code may be used. In some cases we may want to refer to one of the other registers in the CPU. Since in most instances the number of CPU registers is small, part of the operation code bits may be used to address one of these registers. When we come to addressing a word in memory, the number of bits required for this purpose is usually quite large. In general, we require memory arrays of at least a few thousand words, so that both program and data may be conveniently stored. Some examples of the number of bits required to reference an arbitrary location are given below:

For	1024	words of	memory,	10	bits are	required	for reference			
	2048	"	"	"	, 11	"	"	"	"	"
	32,768	"	"	"	, 15	"	"	"	"	"
	65,536	"	"	"	, 16	"	"	"	"	"

Therefore, we may easily require as many as 16 bits for referencing a location in memory. This means that in addition to one word for operation code (opcode), two extra words may be used to express the address relevant to this instruction. The following instruction formats are quite common:

(i) 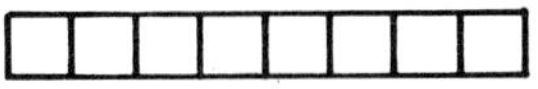

One word (one 8-bit byte); operands are implicit or explicit registers in CPU.

(ii) 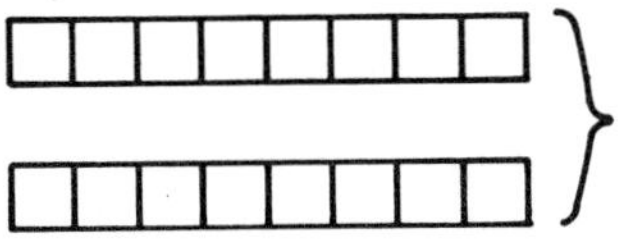

Two words (2 bytes); first word contains the opcode and register references if any, and the second word contains immediate data value.

(iii) 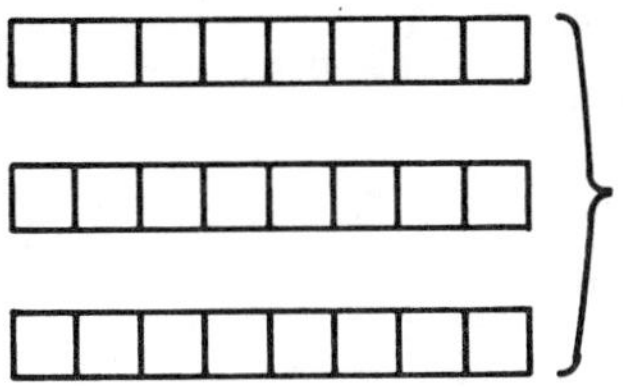

Three words (3 bytes); first word contains the opcode and register references if any, and the next two words contain the address of data item in memory.

Binary operation codes are assigned for a desired set of operations. Based on this assignment, processor logic and hardware is designed to decode and interpret these bit patterns accordingly. At this time, we will not construct a complete list of operation codes, but choose just a few to show the logic of a typical processor operation. Let the partial list of arbitrarily chosen operation codes be as follows:

Operation Code		Assigned Meaning
0011 xxxx xxxx	0001 xxxx xxxx	Load the two words immediately following the instruction as memory address of data to be used later. This 16-bit address will be loaded in a processor register called the data counter (DC).
0000	0001	Load the accumulator (AC) with the data value from the memory location specified by DC
0000	0010	Add to the contents of AC the data value from memory location specified by DC
0000	0011	Store the accumulator contents into the memory location specified by DC

For convenience of representation we will express these codes in hexadecimal, i.e., 31, 01, 02, and 03, respectively. Hexadecimal form is useful in expressing bit patterns. A segment of 4 bits representing binary 0000, 0001, 0010, 0011, 0100, 0101, 0110, 0111, 1000, 1001, 1010, 1011, 1100, 1101, 1110, 1111 is replaced by equivalent hexadecimal digits 0, 1, 2, 3, 4, 5, 6, 7, 8, 9, *A, B, C, D, E. F.* Hexadecimal notation gives a concise and easily manageable representation of large bit patterns. Similarly, all addresses and data values will also be expressed in hexadecimal. Let us assume that a program using the above instructions was written and stored in memory along with relevant data as shown below:

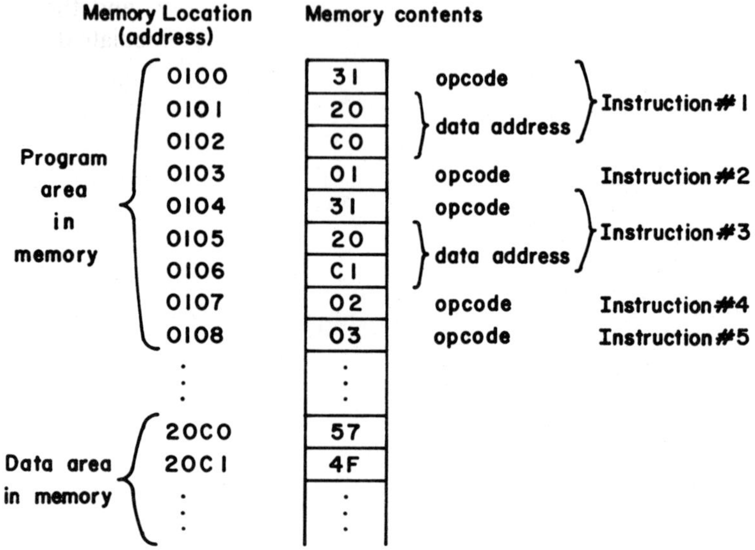

Interpretation and execution of the above program will consist of the following operations:

1. The user (programmer) loads the program counter, PC, with the starting address of the program. The processor registers involved have the following contents:

PC:	0 1 0 0	2 bytes representing the memory address of the first instruction address
AC:	☐☐	1 bytes, may contain any bit pattern
IR:	☐☐	1 bytes, may contain any bit pattern
DC:	☐☐☐☐	2 bytes, may contain any bit pattern

2. When the START button is pressed, the processor moves to a new

state which initiates fetching of the instruction into IR. In actuality this involves two state transitions, the first causing movement of the right half of PC into MA and the second causing movement of the left half of PC into MA and activating the memory readout. The state transition diagram may be drawn as follows:

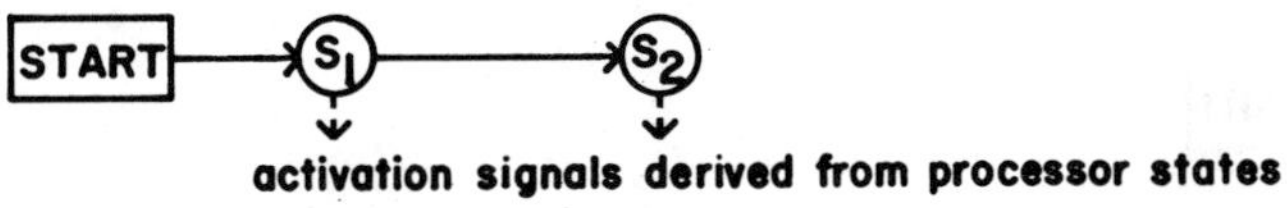

Each state is allowed to last for an interval necessary for hardware operation which is derived from a combination of processor clock pulses. The memory READY line is set to high when readout is complete. If the ready line is high, then a new processor state, S_3, activates the hardware to transfer information from MB to IR, otherwise the processor will go into a wait state, and move into S_3 whenever the ready line is raised, as shown in the following state transition diagram:

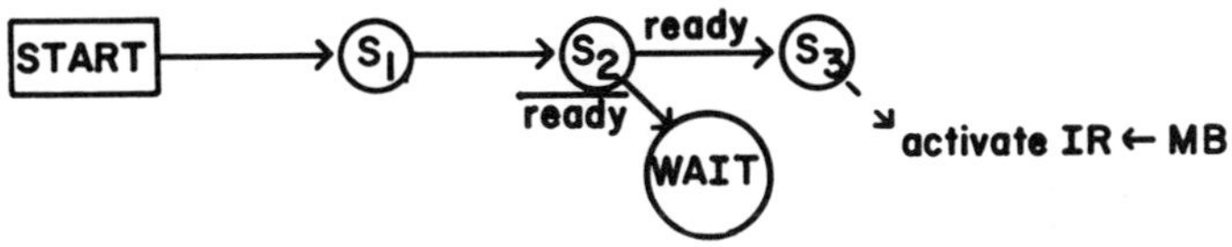

The register contents now appear as follows:

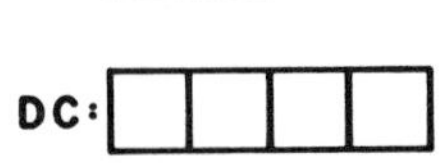

Fetch operation has incremented PC

IR has instruction word just fetched

3. The opcode in IR is now decoded by the processor. In this case interpretation calls for repeating fetch steps twice to get the words at locations 101 and 102, and moving these into DC each time, going

through states S_1, S_2, S_3, S_4 again, as shown in the following state diagram:

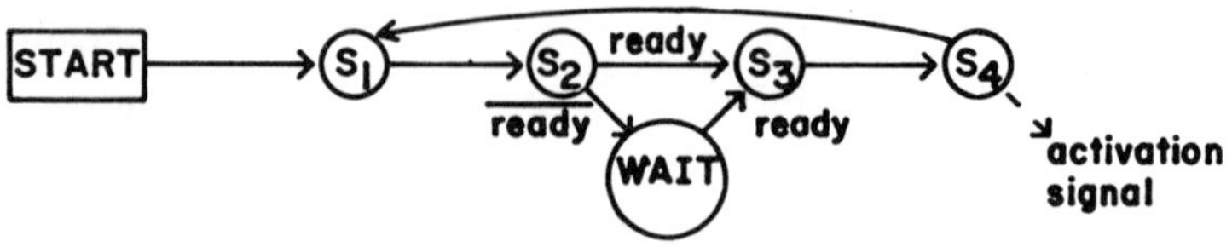

The register contents now appear as follows:

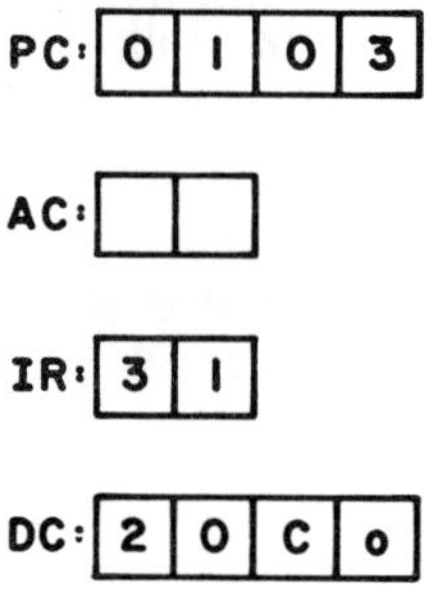

PC incremented twice

Two words of data address fetched from locations 101 and 102

The instruction opcode 31 has now been fully decoded and executed and the processor returns to state S_1.

4. The instruction word for PC value of 103 is fetched, and the register contents change as follows:

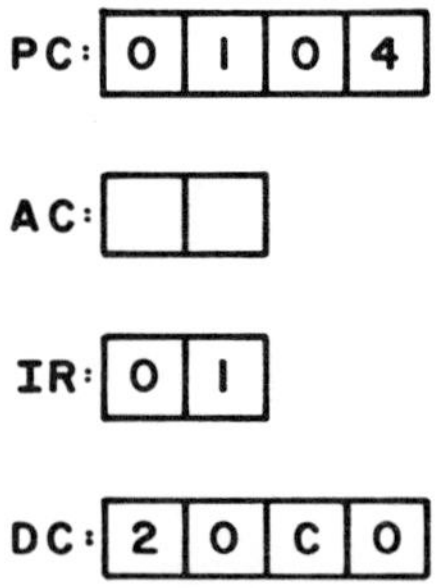

The instruction 01 fetched by going through S_1, S_2, and S_3 of the processor is now decoded and interpreted in states S_4 and S_5 (an additional state S_5 or more may be required for executing some operations depending on their nature and duration). In this case decoding and interpretation requires that data at address 2060 (as pointed to by DC) be loaded into the accumulator, without incrementing PC or DC. The state diagram and resulting changes in registers are shown below:

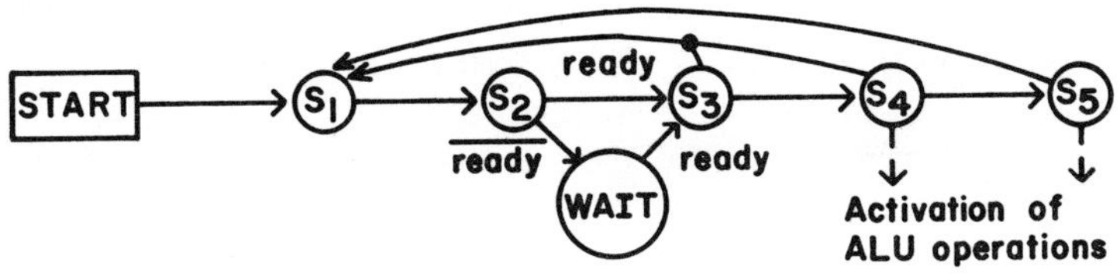

PC: | 0 | 1 | 0 | 4 |

AC: | 5 | 7 |

IR: | 3 | 1 |

DC: | 2 | 0 | C | 0 |

5. Instruction #3 opcode 31 would now be loaded into IR, and PC is incremented. Decoding and interpretation proceeds in a manner identical to Instruction #1 whose opcode was also 31. The register contents at the end of this instruction would be as shown below:

PC: | 0 | 1 | 0 | 7 |

AC: | 5 | 7 |

IR: | 3 | 1 |

DC: | 2 | 0 | C | 1 |

6. Similarly instructions 4 and 5 would be loaded in IR and interpreted

successively. The final register and memory contents are shown below:

PC: 0 1 0 9

AC: A 6

IR: 0 3

DC: 2 0 C 1

Memory locations (addresses)	Memory contents
0100	31
0101	20
0102	C0
0103	01
0104	31
0105	20
0106	C1
0107	02
0108	03
⋮	⋮
20C0	57
20C1	A6

Notice that the data value at location 20C1 has been changed from 4F to A6. This location now has the sum of the data values at 20C0 and 20C1 when program execution was started by pressing the START button. The result shown above has been produced by execution of the program consisting of a specified set of instructions. We could have achieved the same result by writing a somewhat different instruction sequence. For example, rather than specifying the address of the second operand explicitly, we could have reached this address by simply incrementing the DC value corresponding to the first operand. This operation would have required fewer processor state transitions, and thus less execution time. Automatic incrementing of DC is generally not performed because source operands may not always be located in adjacent memory locations. Instruction fetching proceeds in a sequential manner except when a branch operation is performed and in these cases PC would be incremented by a suitable value during execution of the branch instruction.

4.2 BASIC ARITHMETIC AND LOGIC CIRCUITS

In Chapters 2 and 3, combinational and sequential circuit design techniques were considered. In designing circuits to perform specific arithmetic and logic operations, we used NAND gates as the basic building blocks. When the desired functional capabilities of a circuit are rather well defined, and when the operations it performs are commonly used, then this circuit can be constructed as a single module with appropriate input-output connections. A very simple example is a flip-flop. A *J-K* (Master-slave) flip-flop can be constructed by using two 3-input NAND gates, six 2-input NAND gates, and a NAND gate as inverter. Using discrete components (gates) to implement a function may be satisfactory if the function is used rather infrequently. In this case we can wire the circuit together when we need it, dismantle it when we don't need it, and use the gates for other circuits. However, if flip-flops are in common use, they can be produced much more cheaply as a single module in the form of an integrated circuit (IC). This, in fact, is the case.

Many commonly used functions are available as single chip integrated circuits. These ICs range in complexity from as simple as a few gates or flip-flops per chip to a whole processor on a single chip. The integrated circuits are packaged in a suitable ceramic or plastic housing to protect them from damage. Input to these integrated circuits, and output from them are achieved through a number of connectors known as pins. We list below some of the common functions available as integrated circuit packages (IC chips):

1. *Small Scale Integration (SSI) Functions* (a few gates per chip or package)
   Gates (aggregates of NAND, NOR, or inverters), Buffers, Flip-flops, Multivibrators, etc.
2. *Medium Scale Integration (MSI) Functions* (a dozen or so to about one hundred gates per chip or package)
   Adders, Comparators, Complementers, Counters, Shift Registers, Multiplexers, demultiplexers, small size read-write memory or register files, read-only-memories, and a host of other arithmetic and logical functions, e.g. a small ALU, etc.
3. *Large Scale Integration (LSI) Functions* (a few hundred to a few thousand gates per chip or package)
   64-bit or higher read-write memories (RAM), read only memories (ROM), programmable read-only memories (PROM), arithmetic and logic units (ALU), processors, i.e., a CPU on a single chip known as microprocessor, I/O interfaces, peripheral device controllers, etc.
4. *Very Large Scale Integration (VLSI) Functions* (a few thousand to several hundred thousand gates per chip or package)
   Several hundred kilobit memories, 16-bit, 32-bit or 64-bit processor on a single chip, high performance processors with redundant circuits

for military and space applications, video telephone circuits and other communication system functions.

In designing arithmetic and logic circuits for an arithmetic and logic unit (ALU) of a digital computer or other specific applications, IC packages rather than discrete gates can be conveniently employed. This saves cost, simplifies wiring, and reduces physical space and power consumption requirements drastically. Integrated circuit characteristics such as propagation delay, power dissipation, and fanout depend on fabrication techniques used in their production. For example, there are two families of IC packages based on transistor-transistor logic (TTL) known as series 54/74 families of compatible T^2L circuits. Compatibility implies that output levels from any package in the family meet input levels required by any other package. Series 54/74 is characterized by propagation delay per gate of 10 nanoseconds, power dissipation per gate of 10 milliwatts, and operating temperature range of 0–70°C for series 74 and −55 to 124°C (military range) for Series 54. We will now consider some of the basic arithmetic and logic circuits which may be used in the design of ALU.

4.2.1 Addition Circuits

In Chapter 2, we considered single bit half-and full-adder circuits. Suppose now we want to construct an 8-bit adder circuit. Commercially, 2-bit and 4-bit binary full adders are available in Series 74, as listed below:

Name	Number	Add time	Carry time	Power dissipation
		(nano-seconds)	(nano-seconds)	(milliwatts)
1-bit full adder (gated)	7480	52	10.5	105
2-bit full adder	7482	25	14.5	175
4-bit full adder	7483A	16	10	310

In this case 7480 is a full adder with gated complementary inputs, complementary sum outputs, and inverted carry output. We could use any of the above ICs to construct an 8-bit adder. However, the choice of two 4-bit full adders seems to be more appropriate for this purpose from the point of overall add time and power dissipation. Generally, in designing adder circuits we make the carry bit propagate from the least significant addition bits to the most significant. However, for a large number of addition bits, this can cause a large propagation delay in the overall addition process. For example, our 8-bit adder will have two stages:

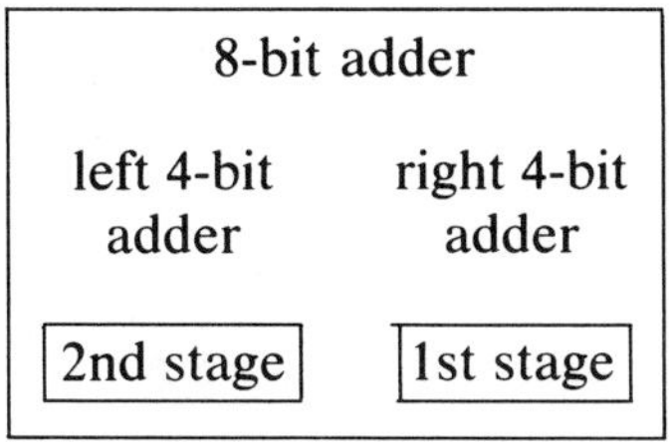

The addition of the left 4-bits (2nd stage) cannot proceed until we have a carry bit from the right 4-bit adder (1st stage). Therefore, adder inputs are fed directly to *carry look-ahead* circuitry which produces the carry from this stage (across 4-bits) even before the addition for this stage has completed. This is accomplished through a carry by-pass circuit included for a group of adder digit positions. Thus, a carry which is generated by a lower order group and which has to be propagated to a higher order group has to pass only one level of logic circuits, instead of being propagated through the individual addition stages of the group. If slower adder operation is acceptable, then we may choose 74LS83 which dissipates much less power, only 75 milliwatts. It is slower (33 and 50 nanoseconds add and carry time, respectively) because the carry bit is serially propagated from the least to the most significant bit of each stage. The pin connections and schematic diagram for 7483A/74LS83 are shown in Fig. 4.15. An 8-bit adder schematic diagram using two of the above IC packages is shown in Fig. 4.16.

4.2.2 Complementation Circuit

The complementation circuit forms an important component in ALU design. In addition to other operations where complementation may be required, we can use the addition circuit designed previously to perform subtraction (using 1's complement arithmatic) simply by complementing the subtrahend and adding it to the minuend as will be explained shortly. The same circuit can be used to perform addition *and* subtraction by putting a true/complement circuit between the addend/subtrahend and the adder. The true/complement circuit can then be signaled to perform complementation of input data or leave it in true binary form depending on whether addition or subtraction is desired. A 4-bit true/complement IC package is shown in Fig. 4.17(a),(b).

In general it is not necessary to show physical placement of leads (pins)

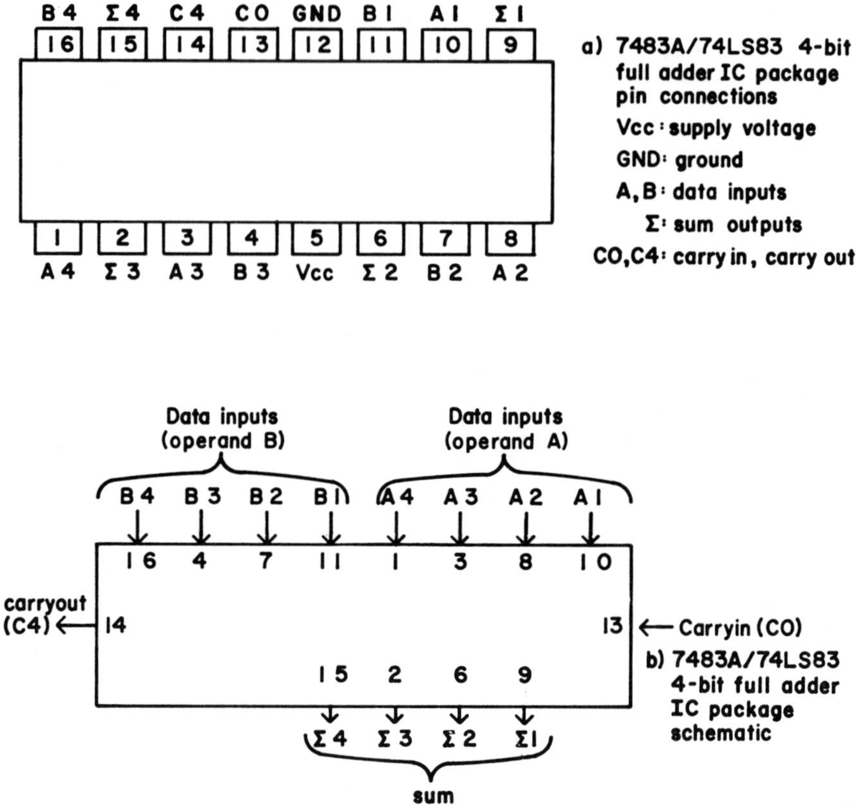

Figure 4.15 (a) Pin connections and (b) Schematic diagram for 7483A/74LS83 IC package containing 4-bit addition circuitry

for ICs when discussing their functional capabilities. Detailed information on these ICs can be found in manufacturer's data books, such as the *TTL Data Book for Design Engineers* published by Texas Instruments. Therefore, we will normally confine ourselves to simply describing functional capabilities with data input-output and control leads.

Returning now to the use of an adder for performing binary subtraction, let us first consider representation of binary numbers in true/complement forms. Using 8-bit representation, all 8-bits will be used but the leftmost bit will be assigned to represent whether the number is positive or negative, it will be 0 for positive numbers, and 1 for negative numbers. If the leftmost bit is zero, then the remaining seven bits represent the number

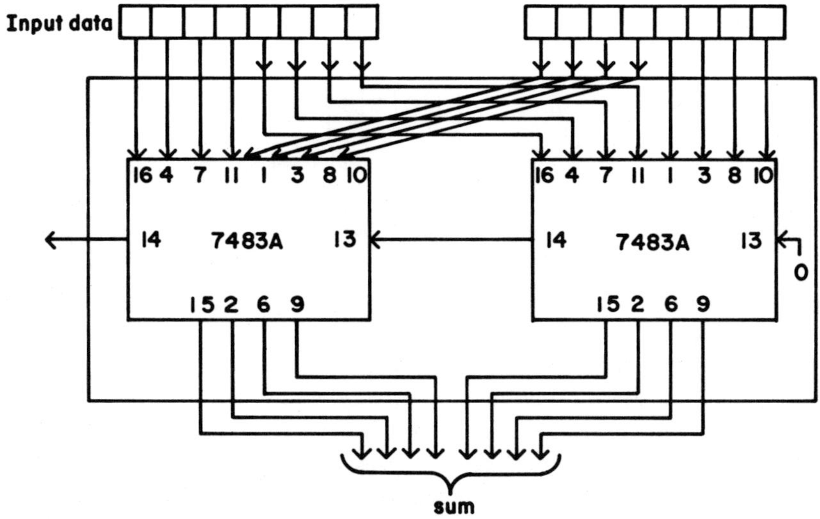

Figure 4.16 An 8-bit adder using two 4-bit adders. Typical hardware characteristics of the 8-bit adder are 23 nanoseconds overall add time, and 620 milliwatts power dissipation using 7483A's

Figure 4.17(a) 4-bit true/ complement, zero/one IC pin connections

A: inputs, Y: outputs
B,C: control inputs
NC: no internal connection

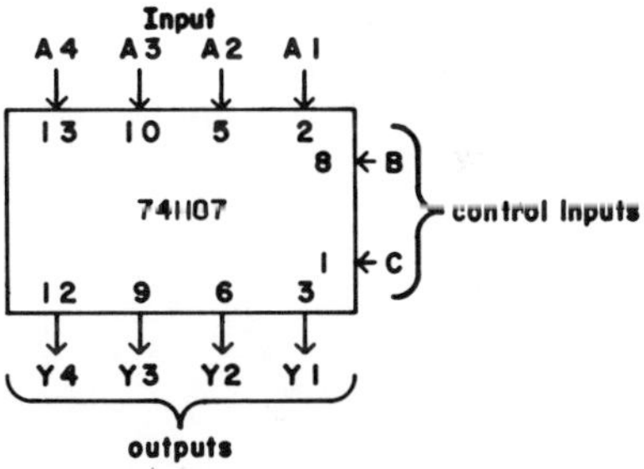

Figure 4.17(b) Schematic diagram for true/ complement circuit

Control,BC:	operation performed
00:	complement
01:	true
10:	all ones
11:	all zeros

in true binary form, otherwise it is in complement binary form. For example,

decimal 25 is represented as 0 0 0 1 1 0 0 1

↓ complement ↓

decimal −25 is represented as 1 1 1 0 0 1 1 0

The complement binary representation shown above is known as the *1's complement*. This form is derived by first representing the magnitude of the number in true binary form; any leading positions available for representing the magnitude but not occupied by the given number are filled with 0. If the given number is positive then a 0 is placed in the sign-bit position. If the number is negative than a 1 is placed in the sign position while every bit of the pattern representing the magnitude is complemented, i.e., 0 is changed to 1 and 1 is changed to 0. Thus we can change the sign of any number by complementing the magnitude bits and also complementing the sign bit. Thus positive and negative numbers can be represented with equal ease and manipulated as such. At the same time, the adder circuit can also be used to perform subtraction by simply complementing subtrahend and performing addition. The requirements for circuits to perform arithmetic operations using 1's complement form can be illustrated by the following examples:

(i) Two positive numbers when added produce a positive result.

(a) +23 : 0 0 0 1 0 1 1 1

+17 : 0 0 0 1 0 0 0 1

+40 : 0 0 1 0 1 0 0 0 All 8 bits added; the result is correct.

(b) +69 : 0 1 0 0 0 1 0 1

+85 : 0 1 0 1 0 1 0 1

-101 : 1 0 0 1 1 0 1 0 — All bits are added, this result is too large ($>2^7$) as indicated by erroneous leftmost bit, i.e. resulting sum has incorrect sign (called overflow)

(ii) Two negative numbers when added produce a negative result.

(a) -23 : 1 1 1 0 1 0 0 0

-17 : 1 1 1 0 1 1 1 0

---

1 | 1 1 0 1 0 1 1 0 — All 8 bits added. Carry out must be added to rightmost position, called end-around-carry, to produce the correct result.

carry out → 1

(b) -40 : 1 1 0 1 0 1 1 1

-69 : 1 0 1 1 1 0 1 0

-85 : 1 0 1 0 1 0 1 0

---

1 | 0 1 1 0 0 1 0 0

carry out → 1

$+101$: 0 1 1 0 0 1 0 1 — Result too small ($<-2^7$) indicated by erroneous leftmost bit, i.e. resulting sum has incorrect sign.

(iii) Operands with opposite signs produce a positive or negative result depending on whichever is larger. (The magnitude of sum can never be too large).

(a)

23	:		0 0 0 1 0 1 1 1
−17	:		1 1 1 0 1 1 1 0
		1	0 0 0 0 0 1 0 1
			1
+6	:		0 0 0 0 0 1 1 0

(b)

−23	:	1 1 1 0 1 0 0 0
+17	:	0 0 0 1 0 0 0 1
−6	:	1 1 1 1 1 0 0 1

Thus the adder can be used to perform subtraction simply by taking the 1's complement of the subtrahend (including the leftmost sign bit) and adding it to the minuend. A combined 8-bit adder/subtractor circuit realization is shown in Fig. 4.18. We have so far discussed the realization of adder and complementer circuits, including the use of a complementer circuit to allow subtraction operation on the adder. The arithmetic and logic unit of a digital computer includes facilities for a large number of other operations. Some of the other common operations are shifting the contents of the accumulator (or other registers); determining whether the accumulator has a positive, negative, or zero value; comparing the contents of two registers, i.e., A and B to determine whether $A > B$, $A = B$ or $A < B$. Some of these operations require additional hardware, others can be derived from the hardware which is already available. Design of shift registers was considered in detail in Chapter 3. At that time we also considered some integrated circuit packages available for this purpose.

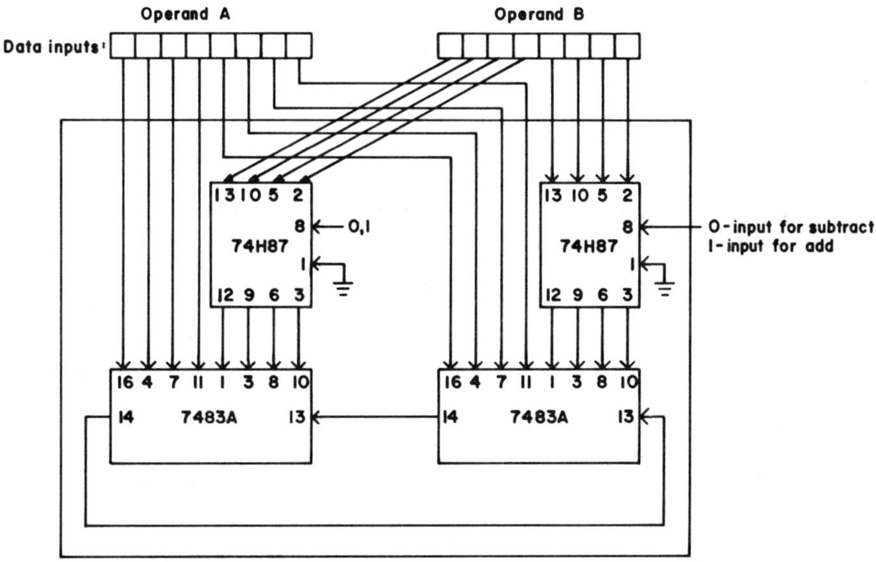

Figure 4.18 8-bit Adder/Subtractor circuit. Level at input lead (pin 8) enables addition or subtraction operation as desired

If the shift operations only involve the accumulator register, then this register itself can be constructed as a shift register. On the other hand, if shifting is to be provided for a large number of registers, then we may have a shift register internal to the processor to which source data would be routed for shift operations. In either case, the implementation is fairly straight-forward and we will not further elaborate on this, at this point.

Arithmetic comparison of *A* and *B* can be performed by subtracting *B* from *A* and checking the difference. For logical (bit by bit) comparison we will employ additional circuitry. Standard 4-bit comparators are available and they can be easily configured for 8-bit comparison.

Since the operations which the arithmetic and logic unit of a computer are required to perform are fairly standard, it is generally easy to find a single integrated circuit package which contains most of these functions. Appropriate data inputs and control signals can then be provided to this IC package to perform one operation or a sequence of operations. Examples of 4-bit ALU integrated circuit packages are Texas Instruments 74181 and Fairchild Semiconductor 9340. Both of these are high-speed TTL/MSI devices. We will discuss the first one in detail in the following section to see how it can be incorporated in the design of a processor.

4.2.3 Arithmetic and Logic Unit

The integrated circuit package 74181 can be operated in two modes: arithmetic and logic. In each mode it can perform sixteen operations. In the arithmetic mode it can perform addition, subtraction, shifting one of the operands by one bit, magnitude comparison, and twelve other arithmetic operations. In the logic mode it can perform exclusive—OR, comparison, AND, NAND, OR, NOR, and ten other logic operations. Fig. 4.19 shows a schematic diagram for this integrated circuit package.

As indicated in Fig. 4.19, a 4-bit code at the input leads ($S4,S3,S2,S1$) selects the operation to be performed. The selected operation is performed by using two 4-bit data inputs (B,A) producing a 4-bit data output (Z). Arithmetic operations may produce a result in excess of 4-bits, causing a carry output (C_{n+4}) which may be used as the carry input (C_n) to the next stage for operands containing more than 4-bits. The connections C_n and C_{n+4} serve as ripple-carry input and output for the device. For high speed operations involving operands more than 8-bits long, provision has been made for simultaneous carry generation via cascading outputs, p and g which are connected to look-ahead carry generator circuits, such as 74182 (see section 7.1).

For arithmetic operations (internal carries enable), the input C_n should be high for 4-bit operations. For operations involving more than four bits this will correspond to the rightmost four bits (least significant) of the

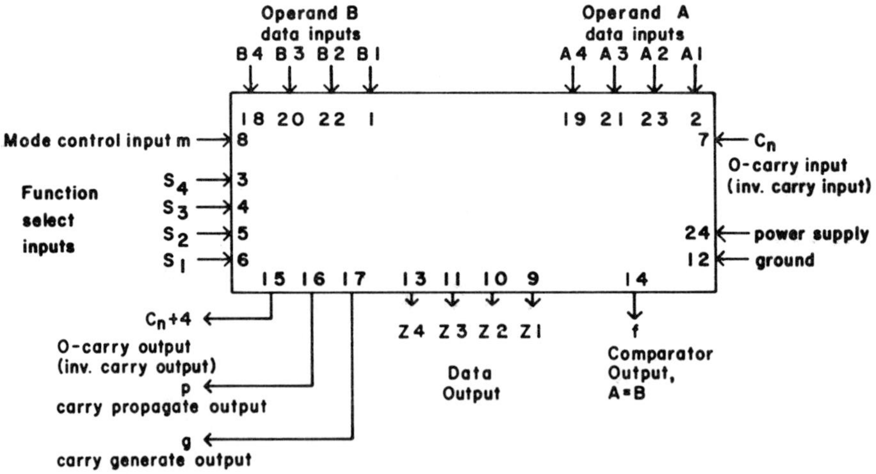

Figure 4.19 Integrated circuit package for 16 possible arithmetic (m = 0) and 16 possible logical (m = 1) operations on two 4-bit operands; low level for m enables internal carries required for arithmetic operations

Selection (operation codes)				Logical operations ($M = H$)		Arithmetic operations ($M = L$; $C_n = H$ means no carry in, $C_n = L$ a carry in & output is as shown plus 1 in this case)	
S_4	S_3	S_2	S_1	Name	Output, Z_i	Name	Output (including carryout)
0	0	0	0	Not A	A_i'		A
0	0	0	1	NOR	$(A_i + B_i)'$		$A + B$
0	0	1	0		$A_i'B$		$A + B'$
0	0	1	1	0	0		2 complement of -1
0	1	0	0	NAND	$(A_iB_i)'$		A plus AB' (Boolean)
0	1	0	1	Not B	B_i'		$(A + B)$ Plus AB'
0	1	1	0	Exclusive -OR	$A_i'B_i + A_iB_i'$	Subtract (without end around carry)	A Minus B minus 1
0	1	1	1		A_iB_i'		AB' Minus 1
1	0	0	0		$A_i' + B_i$		A Plus AB
1	0	0	1	EQUAL	$A_i'B_i' + A_iB_i$	Add	A Plus B
1	0	1	0	B	B_i		$(A + B')$ Plus AB
1	0	1	1	AND	A_iB_i		AB Minus 1
1	1	0	0	1	1		A Plus A(1 left shift)
1	1	0	1		$A_i + B_i'$		$(A + B)$ Plus A
1	1	1	0	OR	$A_i + B_i$		$(A + B')$ Plus A
1	1	1	1	A	A_i		A Minus 1

Table 4.1 Functions available on 74181 arithmetic and logic IC package

operands. By activating the subtraction operation we can perform comparison of data values on inputs A and B. Since the subtraction is performed by the 1's complement method, the carry out C_{n+4} will be low (indicating a carry) if $A < B$, C_{n+4} will be high if $A \geq B$. An additional output, f, is generated internally from the Z outputs to indicate whether $A = B$ (f high) or $A \neq B$ (f low).

Table 4.1 gives a summary of the operation codes and the corresponding operations performed by 74181. An examination of Table 4.1 will reveal that the 74181 package provides most commonly required arithmetic and logical operations directly. For subtraction an end around carry must be incorporated.

4.3 ARITHMETIC AND LOGIC UNIT ORGANIZATION

In the preceding section, we discussed the design of arithmetic and logic circuits to perform operations on fixed-length operands. Typically, a computer program would include a sequence of instructions, and each instruction will invoke some ALU operation(s). Quite often, we want to check the state of the operation just performed to decide which instruction of the sequence will be executed next. For example, we may want to check whether the operation just performed resulted in a *zero* value in the accumulator, or caused an incorrect value in the accumulator because of an erroneous *carry* (see the preceding section on arithmetic addition and subtraction). We may also be interested in the *sign* of the result produced by the operation just performed. The eight operand bits can contain a binary value or binary-coded-decimal (BCD) value using 4-bits to code each decimal digit. Arithmetic operations can be performed directly on BCD data. However, we will have to keep track of the carry out of half word (4-bits) rather than the full word (8-bits) as in the case of binary arithmetic. We may call this *auxiliary carry* or *half carry*. The data stored in memory may contain character information which conforms to even *parity* (even number of 1s in the pattern representing the character) or odd parity (odd number of 1s in the pattern representing the character). In moving this information from memory to ALU, a *parity* bit may be set appropriately. Therefore, in designing an ALU, a set of *status flags* are included to indicate the conditions associated with ALU operations. We can designate one 8-bit register in the ALU as the status register; the individual bits can represent the on/off conditions of the corresponding flags. A typical configuration of status register is shown below:

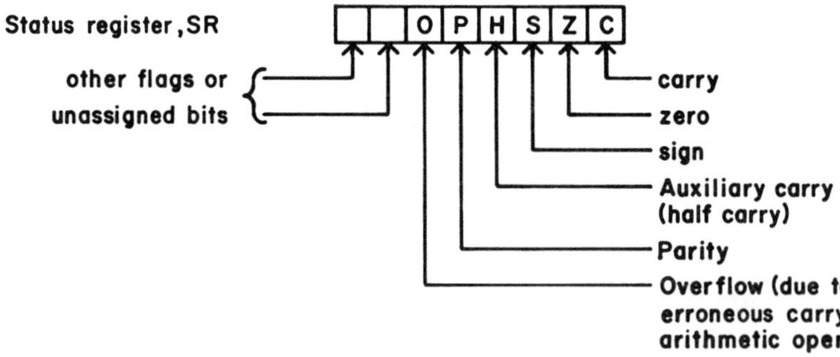

The arithmetic and logic unit requires one or two source operands for any one of the operations. If the operation requires a single operand, then this operand may be contained in the accumulator. However, for two operands we need an additional register to hold source data. This register can also serve as the temporary destination for results produced by the ALU before routing to another register or memory location. For unary operations, we may again use this register as a temporary holding area, and perform ALU operations without disturbing the contents of the accumulator. The central processing unit (ALU + Controller) of a computer contains a number of other registers which have data paths to or from the ALU.

Figure 4.20 shows the sets of registers associated with the ALU in two of the currently popular microprocessor units.

In addition to those registers which were discussed before, there are two registers–an index register (IX) in MC6800, and a stack pointer in MC6800 as well as Intel 8080. When the indexing option is used in memory reference instructions, then the one byte displacement value contained in those instructions is added to the contents of the index register to calculate the effective address of the operands. This way the index register can be used to point to the base of a table and the displacement value as

(a) *Motorola MC68000 Microprocessor Registers*

(b) *Intel 8080 Microprocessor Registers*

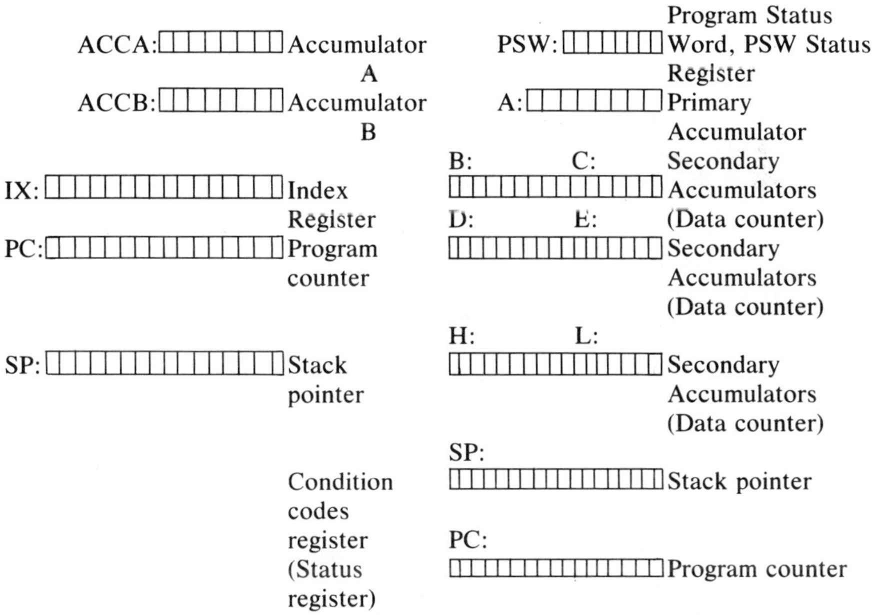

Figure 4.20 Examples of processor registers associated with ALU operations

indices, ranging from 0 to 255, to individual entries of the table. We may also visualize the index register as a page register where each page is 256 words long. In this case the instruction register (IR) need only contain the address of a page. With 8 address bits in IR, 256 pages can be referred. Including one bit for indexing, a total of 9 bits in IR can refer to any one of 256 × 256 memory locations–a range of 65,536 addresses which would require 16 bits in the IR for direct reference.

The stack pointer (SP) is used as a pointer to an area in memory (RAM) which contains elements of a stack. The stack is useful for temporary storage and retrieval of successive words of information. This information is stored or retrieved without having to specify an address. A store operation on the stack results in storing the information on top of the stack, and a retrieval operation results in retrieving information from the top of the stack. The stack pointer value is adjusted automatically in both cases according to the operation performed.

As an example, consider a portion of RAM as shown in Fig. 4.21.

Now, if a word (byte) of information is stored in the stack, it is stored at location 0253 (hexadecimal) or 0000001001010011 (binary), the address contained in SP; and SP is decremented by one automatically immediately after the storage operation. The new SP value is 0252 (hex). Conversely, when an instruction for retrieving information from the stack is executed, the SP value is incremented first and then the item now pointed to by SP is retrieved. Because of this last-in-first-out (LIFO) operation, the stack serves a very useful purpose in processor operations. For example, in case of an interrupt, the current status (information) of the processor can be stored, and then retrieved subsequently when the interrupt has been serviced. Similarly, in making calls to a subroutine, the return address

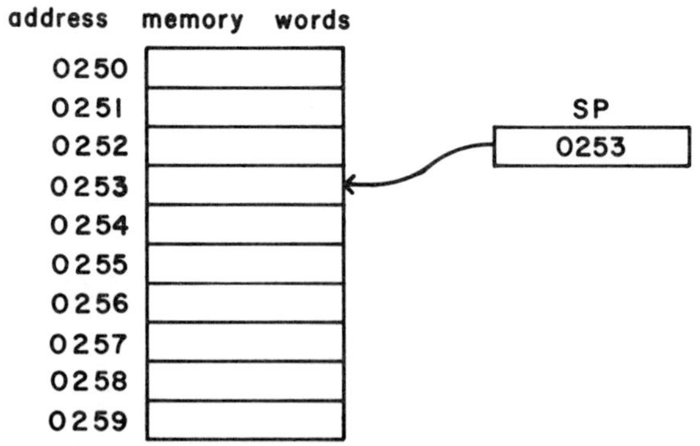

Figure 4.21 A stack declared in RAM, with SP pointing to top of stack

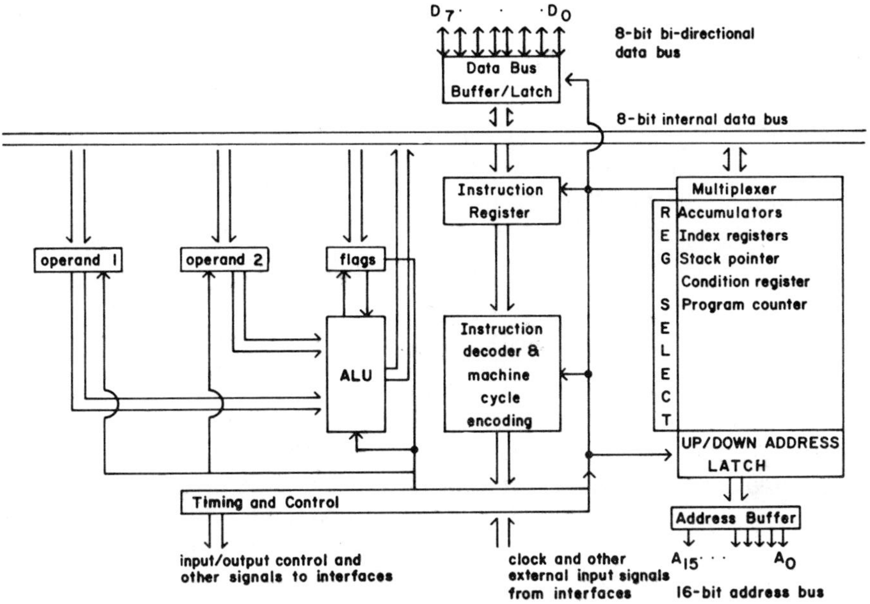

Figure 4.22 ALU operating environment which constitutes a processing unit. The processing unit includes, in addition to ALU, a register array and logic to route data to or from these registers; instruction registers and decoder/interpreter hardware; ALU operands and flag registers; timing and control for internal/external operations; data bus for input/output; address output bus and an internal bus for routing of information

can be stored on the stack. For nested calls, we can push successive addresses on top of the stack, and upon return pop these addresses one by one because we always want to return to the most recent return address first. The stack may also be used to store other data items as deemed necessary or convenient by the programmer. Since all references to stack are to the topmost item only, no address calculation is required during instruction decoding and interpretation, thereby reducing instruction execution time.

While some of the registers referred to above may not be considered a part of the ALU organization, they are intimately related to ALU operations particularly when we consider performing sequences of operations on the ALU. As mentioned earlier, a sequence of operations to be performed by the ALU is usually expressed by coded bit patterns known as instructions. A set of instructions constitutes a program. A program may reside in a RAM or ROM. The processing unit provides the address of the instruction to be executed next, this instruction is fetched from memory as data consisting of a binary bit pattern and is loaded into the instruction register. Appropriate circuitry in the processor must now decode the instruction, and activate the ALU circuitry in an appropriate manner according to predefined time intervals. We will discuss the design of decoding circuits in more detail in the next section. However, it is relevant at this time to show the ALU organization within the context of a central processing unit. This is shown in Fig. 4.22.

4.4 DECODER AND DECODING CIRCUITS

Given a bit pattern representing the address of some memory location or an operation code (opcode), we require a circuit to decode this bit pattern to select (energize) the appropriate memory location for storing/retrieving data or to enable the appropriate computer elements to perform the operation indicated by the operation code. The decoding operation may also be required to convert arbitrary bit patterns into desired output patterns such as BCD to decimal, Excess-3 codes into decimal, etc. If an n-bit input pattern represents straight binary information as in the case of memory addresses or operation codes, then the output will consist of 2^n lines, only one of which will be "on" in any given case. The circuit to perform this operation is known as n-to-2^n (or 1 out of 2^n) decoder. A 2-to-4 line decoder may be represented as shown in Fig. 4.23(a). Generally, we will also include another input for strobing or enabling of the circuit to perform the decoding operation, as shown in Fig. 4.23(b). A

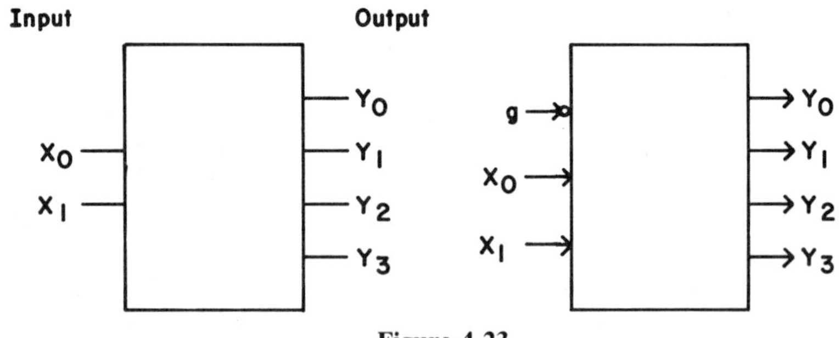

Figure 4.23

(a) 2-to-4 line decoder without enable (strobe) input

(b) 2-to-4 line decoder with enable (strobe) input g. Circuit is enabled when g is low (binary 0)

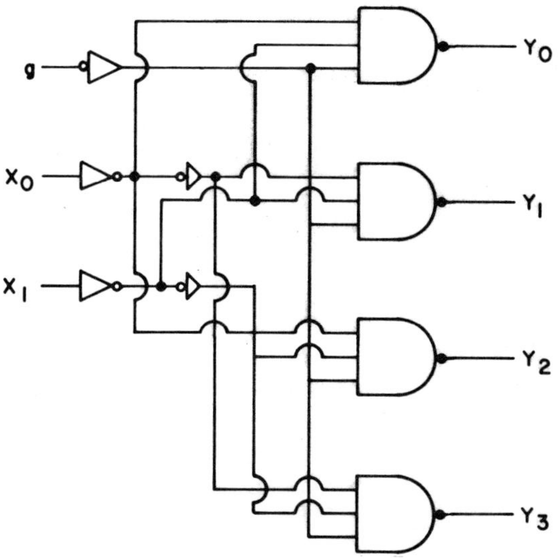

Inputs			Outputs			
Enable	Select					
g	X_1	X_0	y_0	y_1	y_2	y_3
H	X	X	H	H	H	H
L	L	L	L	H	H	H
L	L	H	H	L	H	H
L	H	L	H	H	L	H
L	H	H	H	H	H	H

H-High level; L-low level
x-Don't-care or irrelevant.

Figure 4.24 2-to-4 line decoder

(a) Physical implementation (b) Function table

simple implementation using NAND gates and inverters is shown in Fig. 4.24. In this implementation, when the circuit is enabled (*g* has low level, i.e., is 0), it forces the selected output line to low level. A decoder circuit can also be used to perform demultiplexing or routing of data to one of the selected output lines. This is achieved by using the enable input *g* as a data input line. The decoders/demultiplexers required for common applications need not be constructed with discrete components. They are usually available as standard integrated circuit packages (chips), as shown below:

Application	Available IC
Dual 2-to-4 line decoder Dual 1-to-4 line demultiplexer	SN4155/74155 (Totem pole outputs)
3-to-8 line decoder 1-to-8 line demultiplexer	SN54156/74156 (Open collector outputs)
3-to-8 line decoder 1-to-8 line multiplexer	SN54LS138/74LS138 (Totem pole output)

4-to-16 line decoder	SN54154/74154 (Totem pole output)
1-to-16 line demultiplexer	SN74159 (Open collector output)

Most of these ICs contain features to permit building of larger systems such as 24-line or 32-line decoders without or with only a few extra components. Whether we build decoders by using discrete components directly or ICs, it is useful to examine some of the techniques to develop an appreciation of the nature of design problems.

4.4.1 One-Level Decoder Circuits

Assume that the input is available in the form of $2n$ lines, n for signal bits and the other n for complements. This assumption is usually true because the input bit pattern, whether a memory address, or an opcode, is contained in a register (with one flip flop for each bit of the pattern). Thus a simple decoder may be constructed as shown in Fig. 4.25. Let a be the total number of AND gates and d be the total number of diodes per gate (assuming that AND gates are constructed from simple diode networks, each gate has a minimum of as many diodes as the number of inputs). For the one level realization shown in Fig. 4.25, $a \equiv 2^n$ and $d \equiv n$. The approximate cost of realization is $C \equiv n2^n$. The number of levels (indicative of speed) is $l \equiv 1$. The fan-in (i.e., maximum number of inputs

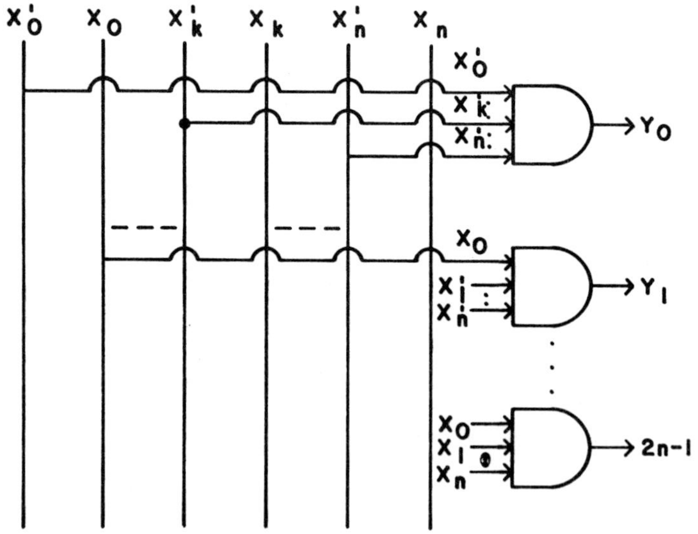

Figure 4.25 One level decoder design

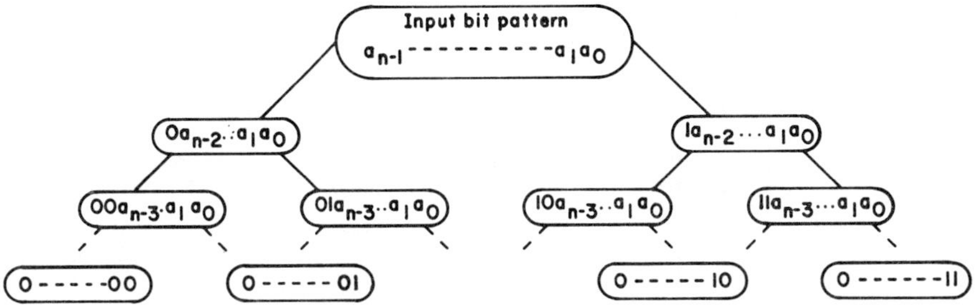

Figure 4.26 Binary tree for decoder design

to a gate) is n: The fan-out is 2 (for input driver, i.e., loading of register flip flops).

4.4.2 Simple Tree Decoder Circuits

A binary tree for an n-to-2^n line decoder can be drawn as shown in Figure 4.26. For an input pattern of n-bits, the tree would have $n - 1$ levels, every gate would have a maximum of 2-inputs, and the output of every gate would serve as one of the inputs to every gate of the successor level. An example of a physical realization using this principle is shown in Figure 4.27. For an input bit pattern of n bits.

$a \equiv 4 + 8 + \text{-------} + 2^n = 4(2^{n-1} - 1)$
$d \equiv 2$
$c \equiv 2 \times 4(2^{n-1} - 1) = 8(2^{n-1} - 1)$
$1 \equiv n - 1$
Fan-in $\equiv 2$
Fan-out $\equiv 2^{n-1}$ (for input drivers); 2(internal)

It is apparent from these expressions that binary tree decoders are very efficient compared to one-level decoders in terms of the number of components (gates) and the fan-in requirements of these components. However, the former contains $n - 1$ levels thereby introducing substantially more delay which may or may not be significant in a given application. Further saving in the number of components (gates) can be achieved by constructing dual binary trees for the decoding operation. For example, to construct an 8-to-256 line decoder, we first visualize two 4-to-16 line decoders which are constructed by using two 2-to-4 line one level decoders each. Outputs of these four 2-to-4 line decoders are combined to produce two 4-to-16 decoders, and the outputs of these two 4-to-16 de-

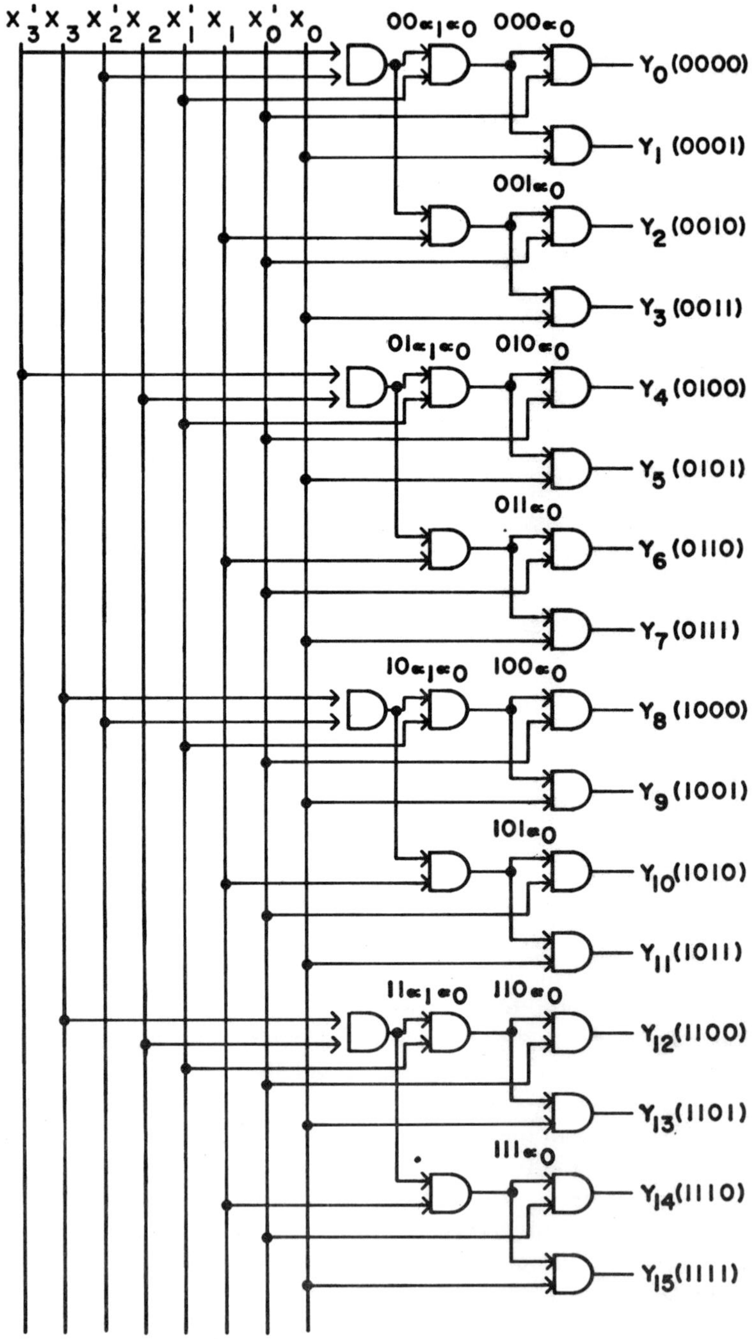

Note:
α denotes value to be assigned to X_2'

Figure 4.27 4-to-16 line tree decoder

coders are combined to produce one 8-to-256 line decoder. In this case the total number of gates used are

(4+4+4+4) +	(16+16) +	(256)
first level	second level	3rd & final level

for a total of 304; thus

$a \equiv 304$
$d \equiv 2$
$c \equiv 608$
$1 \equiv 3$
Fan-in $\equiv 2$
Fan-out $\equiv 2$ (for input drivers)

Of course, the internal fan-out for gates is much larger in this case because of the manner in which outputs from one stage serve as inputs to successor stage. The general expressions for n-to-2^n line decoders using the dual-tree approach can be easily derived as follows:

$$a \equiv n \sum_{i=1}^{l} \frac{2^{2^i}}{2^i}$$

$d \equiv 2$
$c \equiv 2a$
$1 \equiv \lceil \log_2 n$
Fan-in $\equiv 2$
Fan-out $\equiv 2$ (for input drivers); $2\lceil n \div 2$ (internal)
where $\lceil$ denotes ceiling operation, i.e. next higher integer value

The preceding examples have shown how discrete components (gates) can be used to construct decoders for small or large number of inputs. In practice, decoders for small number of inputs are available commercially and they are quite inexpensive. For large number of inputs, decoders are constructed by using these ICs rather than discrete components. For example, a 6-to-64 line decoder system can be built by using nine 3-to-8 line decoders as shown in Fig. 4.28. This technique also permits us to distribute decoding operations to different locations in the system in a manner which is similar to telephone line operations or mail delivery using postal codes.

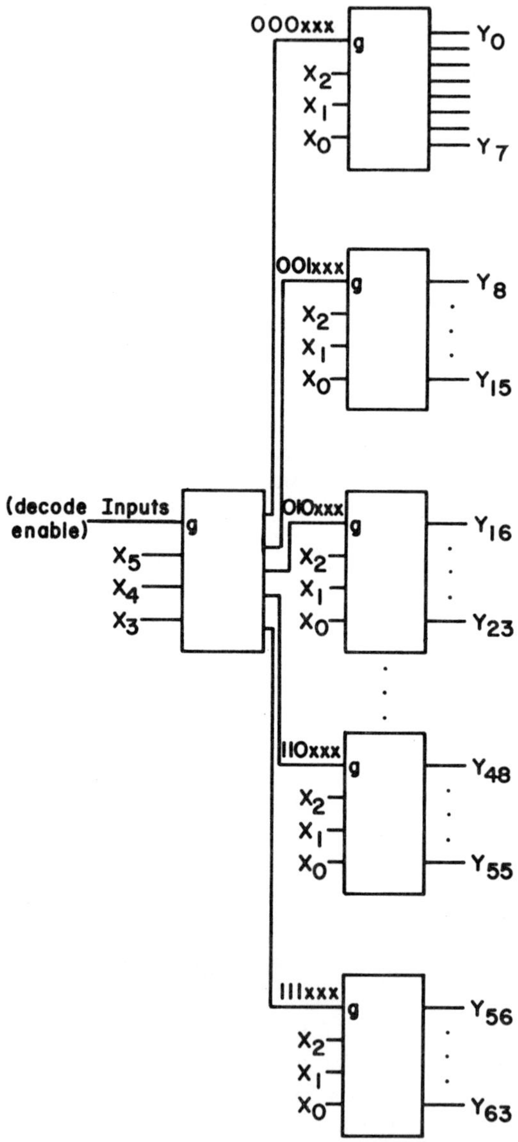

Figure 4.28 6-to-64 line decoder system using 3-to-8 line decoder ICs

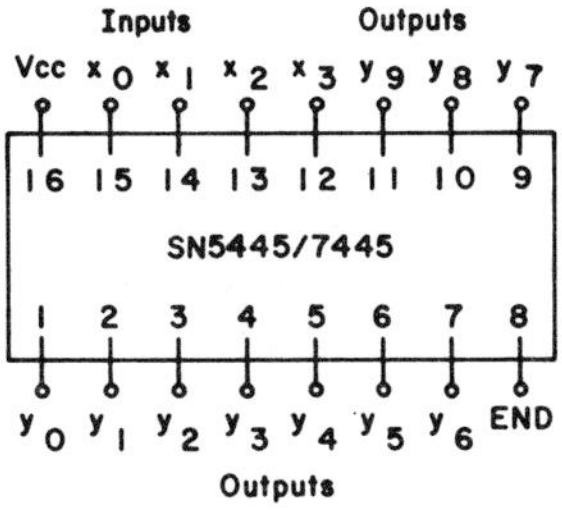

(a)

No	Inputs				Outputs									
	x_3	x_2	x_1	x_0	y_0	y_1	y_2	y_3	y_4	y_5	y_6	y_7	y_8	y_9
0	L	L	L	L	L	H	H	H	H	H	H	H	H	H
1	L	L	L	H	H	L	H	H	H	H	H	H	H	H
2	L	L	H	L	H	H	L	H	H	H	H	H	H	H
3	L	L	H	H	H	H	H	L	H	H	H	H	H	H
4	L	H	L	L	H	H	H	H	L	H	H	H	H	H
5	L	H	L	H	H	H	H	H	H	L	H	H	H	H
6	L	H	H	L	H	H	H	H	H	H	L	H	H	H
7	L	H	H	H	H	H	H	H	H	H	H	L	H	H
8	H	L	L	L	H	H	H	H	H	H	H	H	L	H
9	H	L	L	H	H	H	H	H	H	H	H	H	H	L
I	H	L	H	L	All High									
n	H	L	H	H										
v	H	H	L	L										
a	H	H	L	H										
l	H	H	H	L										
i	H	H	H	H										
d														

(b)

Figure 4.29 BCD-to-decimal decoder/driver

(a) IC package layout
(b) Function table

4.4.3 Decoder Applications

So far we have discussed binary decoders generally applicable in memory addressing applications, opcode decoding, or routing of binary data. There are many other decoder designs such as BCD input to decimal output, BCD input to seven segment output, etc. These decoders find applications in the design of counters, analog-to-digital data conversion, etc. In many instances these decoders which are available as IC's are designed so as to perform the dual function of decoders as well as drivers (providing adequate power) for display devices such as cold cathode indicator tubes, relays, light emitting diodes (LED's), or drivers for MOS integrated circuits. A BCD-to-decimal decoder/driver IC is shown in Fig. 4.29. This decoder accepts normal TTL inputs, and the outputs are capable of driving lamps, relays, or other logic circuits. Similarly, BCD-to-seven-segment decoder/driver ICs are also available for direct applications without requiring extra components even for operations such as suppression of leading/trailing zeros in displays, testing of display lamps, or inhibiting the outputs. An example of an available IC is SN5446/7446. The reader is advised to examine the data sheet for this IC to familiarize himself or herself with the device operation. The standard designations of seven segments and corresponding numerical digit configurations are shown below:

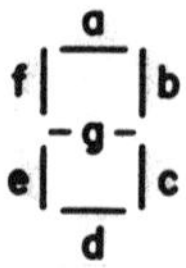

segment designations

Displayed segments	Numeric digits		
 (may be suppressed)	0	Displayed segments (outputs true) ;	a,b,c,d,e,f
	1,	"	: b,c
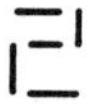	2,	"	: a,b,d,e,g
	3,	"	: a,b,c,d,g
	4,	"	: b,c,f,g
	5,	"	: a,c,d,f,g
	6,	"	: c,e,f,g
	7,	"	: a,b,c
	8,	"	: all seven
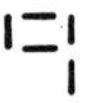	9,	"	: a,b,c,f,g
	10, (invalid)	"	: d,e,g
	11, (invalid)	"	: c,d,g
	12, (invalid)	"	: a,f,g
	13, (invalid)	"	: a,d,f,g
	14, (invalid)	"	: d,e,f,g

blank	15, (invalid)	"	: none

(a blank will also be displayed where zeros are suppressed)

We can conceive a BCD-to-seven-segment decoder as a code converter which accepts four bit input codes and produces seven bit output codes. Decoders of this type can also be conceived as a read-only-memory (ROM) where inputs serve as addresses of memory locations and outputs as the contents of those memory locations.

4.5 STORAGE(MEMORY) CIRCUITS

Memory or storage of information to be retrieved for subsequent uses forms an essential component of computer operation. We have already discussed instances of elementary memory elements. For example, an intrinsic characteristic of sequential circuits is that they use the previous state and current input(s) to produce next state and current output(s). Flip-flops were used to retain these memory states for subsequent use. Thus a flip-flop is the most elementary form of storage circuit and a single flip-flop is capable of storing one bit of information which is either 0 or 1 at any given time. We have also discussed how a number of these flip-flops are put together to form a device capable of storing an arbitrary number of bits of information. These devices were called registers. The number of bits in a *register* define its *length* attribute. Typical lengths of registers used in computer operations are 2, 4, 8, 16, 24, 32, etc. Registers may also have other lengths. Lengths which are multiples of eight are more common because they directly relate to another unit of information known as a byte. A *byte* is generally defined to be eight bits of information although there are instances of computers where bytes consist of six bits of information. While we have defined a register of information, most computers use many such registers. In this case an additional attribute known as the *register address* must also be introduced. If we have, for example, 16 registers we need four bits of information to specify the address of any one of the 16 registers. These address bits are generally contained in a register called the *address register*. Collectively, a set of registers is conceived of as a *Storage* or *Memory* device with two basic attributes—*storage capacity* in terms of *number of words* (or bytes) of information which corresponds to the number of registers, and *word* (byte) *length* corresponding to the register length as defined above. The storage device may have a special name such as a *stack* because of implied addressing capability. The most common form of storage device used as primary memory in computers is known as random-access-memory (RAM) because this device is capable of receiving (storing) information in an arbitrary (random) memory address (location) and likewise retrieving (recalling) information from an arbitrary (random) memory address (location) and the access time is independent of the address being accessed. While conceptually there is no difference between a *register* and *memory word,* there are practical considerations which tend to distinguish these two names. The name *register* is usually used for those storage elements whose output lines have direct connections (or an appearance thereof) to arithmetic and logic circuits for operations such as add, subtract, shift, compare, etc. Likewise, output lines of arithmetic and logic circuits are directly connected to the registers. Addressing in this case implies that selected (addressed) register inputs and outputs are appropriately enabled to participate in the desired operation. However, when the requirements for registers (of information) demand a large number of words, it is impractical to build them all having direct connections to arithmetic and logic circuits of the computer. In this case, we associate a buffer register with the memory device known as *memory buffer register* (MBR). The information (data) to be stored in memory is put in MBR, the address of the location in which the information is to be stored is put in a memory address register (MAR), and the memory is enabled to perform writing (storing) of data in the specified location. For retrieving (read), the address is put in MAR, and the memory is enabled to perform reading, thus bringing information into MBR. Therefore, we separate the internal design of memory circuits from the outside environment by providing suitable means of communication with memory regardless of its internal operation as shown in Fig. 4.30.

The memory buffer register (MBR) contains at least as many bits as the number of bits/word of memory. The memory address register (MAR) contains at least $\log_2 n$ bits for n words of memory capacity. Thus for 2048 8-bit words, MBR has 8 bits and MAR should have at least 11 bits. However, in many designs, a larger MAR, say 16 bits, is used to accommodate

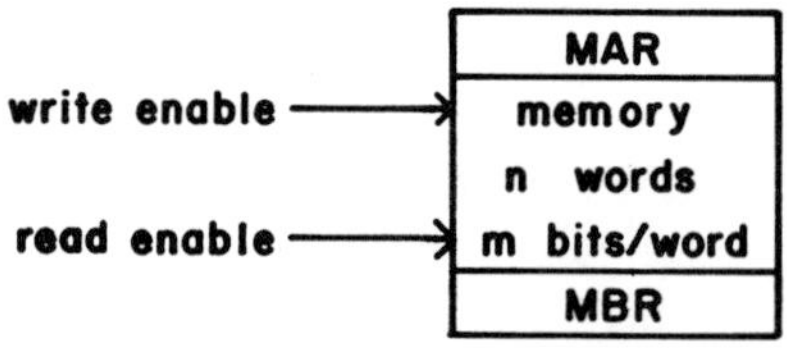

Figure 4.30 Logical organization of memory device

expansion in memory capacity. The unused bits on the left can be kept at zero.

4.5.1 Storage Material and Techniques

Bits of information can be stored on different materials, and many different techniques using the same material. However, in this section, we will confine our discussion to materials and techniques commonly used in the design of random-access-memory (RAM) used for writing/reading of information during computer operations and read-only-memory (ROM) containing information of "permanent" nature which can be accessed randomly but would only permit read operation. Memory devices for these applications are generally fabricated from magnetic or semiconductor materials. In the case of magnetic materials, the polarity of magnetization determines the value 0 or 1 of one bit of information. In the case of semiconductor materials, the signal level (low or high) at the output of a transistor (gate) determines the value 0 or 1 of one bit of information. A basic difference between the magnetic and semiconductor storage devices is that information stored in magnetic materials is non-volatile, i.e., it does not require energy for retaining stored information whereas semiconductor circuits must remain connected to a power supply to retain stored information. On the other hand, reading stored information from magnetic devices generally destroys that information and, therefore, a read cycle must be followed by a write cycle. Reading of information from semiconductor devices is generally non-destructive. To appreciate these and other characteristics we will now examine storage cells each used to store one bit of information, as shown in Fig. 4.31.

The storage cells shown in Fig. 4.31 could well be designed in a simpler manner. For example, in magnetic core storage we have used two wires (X, Y), and collectively these wires are made to carry enough current to cause switching of core polarity when the core is selected. This is called the *coincident-current selection* technique. We could use a single wire and send enough current on this wire to cause switching of the core when the core is selected. This is the *linear selection* technique. We have shown the coincident-current-selection technique because it is more common, and the reasons for this will become clear in the design of memory arrays containing large numbers of cells. The same remarks apply to semiconductor storage cell shown in Fig. 4.31. We have shown a magnetic core as a basic cell for storing information using magnetic material. Many other variations are possible. Similarly, there are many variations on the multiple-emitter transistors shown for the semiconductor storage cell. We

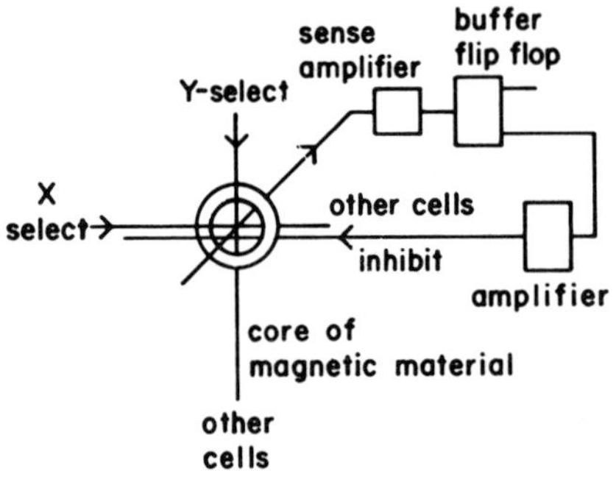

X, Y wires carry current to select a core by writing 0 for reading operation followed by restoring the value of 1 if sensed during read. The wires also carry current to write 1 or 0 in the selected core. An amplifier is required to raise the signal level on the sense wire and store the sensed information by setting or resetting of the buffer flip flop. For reading, the selected core receives current on both X, Y wires and their sum causes switching of the core to 0 polarity. If the core previously had a 0 polarity, no change in flux occurs and the sense amplifier when enabled receives no signal, and its zero output clear the buffer flip flop. If the core previously had 1 polarity, writing of 0 causes flux change giving high amplifier output and thus setting buffer flip flop to 1. In either case read operation is followed by writing of 1 but during this time sense amplifier is disabled and inhibit amplifier is enabled. Thus a 1 is written provided the core had 1 polarity when read was started.

(a)

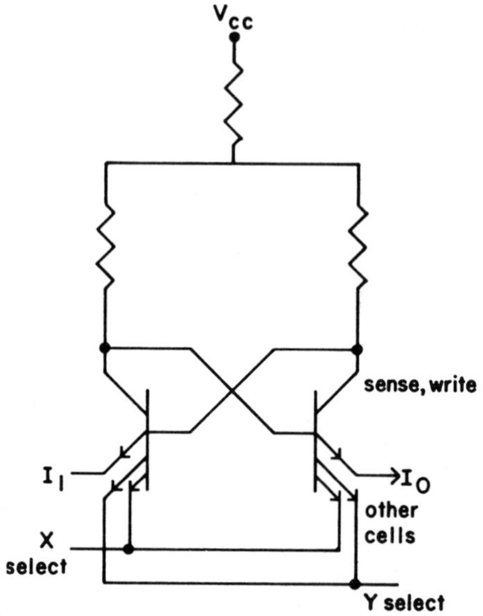

X, Y wires (lines) are normally held low thus providing a path of current for the conducting transistor. Raising X, Y high causes this current to be transferred to the sense line. Thus if the left transistor is conducting (its collector is low) then selection of this cell causes the current to appear on the I_0 sense line. When X, Y lines return to their normal level, the circuit returns to its previous condition. Thus the read operation is nondestructive. To write, the cell is selected, and the write lines I_0 or I_1 are raised as the case may be. This causes the transistor connected to this line to be cut off and the other transistor to conduct. When the select and write lines return to their normal values, the cell will remain in the state caused by the signal on the write line. Depending on the technology or levels of signals we will have corresponding amplifier or read/write circuits associated with the lines.

(b)

Figure 4.31 One cell of storage using (a) Magnetic core technology (b) Semiconductor technology will discuss some of these variations in later sections and also provide reasons for their use.

4.5.2 Memory Arrays

Using the basic memory cell discussed in the preceding section, we can organize memory arrays to serve as primary random access memory (RAM), for computer programs and data. Typically, these memory arrays or storage devices are available in modules of 2048(2K), 4096(4K), 8192(8K), 16384(16K), 32768(32K), 65526(65K also referred to as 64K), etc., bits of read-write storage capacity. The read-write operations are usually assumed to access one full word i.e., all bits of a word at the specified address. We may, therefore, conceive memory modules as consisting of a stack of planes; there being as many planes as the number of bits per word. The number of bits-per-plane then corresponds to the storage capacity of the module in words. Thus for eight bit words, the modules listed above will have capacities of 256(¼K), 512(½K), 1024(1K), 2048(2K), 4096(4K), 8192(8K), etc., words. These modules can also be assembled to provide larger capacity of storage as well as 8, 16, 24, or 32 bits per words of storage. For semiconductor memory, smaller modules are produced as single integrated circuit chips, and these chips are assembled on a single circuit board to provide the desired size random access memory. The semiconductor random access memory is volatile i.e., information stored is lost when power is turned off. For permanent random access storage, we use the read-only-memory (ROM). Information in ROM is stored by a burning-in process (altering circuit contacts mechanically or by electric currents to provide permanent bit patterns) or through a special programming device for programmable read-only-memory (PROM). Examples of programs for which permanent storage is desirable include monitor routines, device controller routines (device drivers), interpreters, and compilers, etc. Some of these programs are required for operating computer hardware and others such as interpreters and compilers are needed to translate user programs written in problem oriented languages into a form acceptable to the processing unit of the computer (see Sections 1.2 and 1.4).

As the requirements for permanent storage increase, it may not be feasible to store all the information in ROM or PROM. In this case we can use auxiliary storage devices such as magnetic tapes or discs. For example, a typical microcomputer system may have 32K bytes of RAM, 8 to 16K bytes of ROM and 200–400K bytes of on-line storage using a small flexible disc commonly known as a floppy disc. This represents a two-level storage hierarchy. Any of the program or data files (named aggregates of information) stored on disc can be accessed at random and read sequentially into RAM when required. New and updated program or data files can be stored from RAM onto disc for later uses. A library of discs can serve as an inexpensive means of off-line storage. Auxiliary storage will be further discussed in Chapter 6.

PROBLEMS

4.1 The accumulator (AC) in ALU usually serves as a register for receiving results of operation such as addition, subtraction, etc. For binary operations, two source operands are required. Assume that one of the operands is in AC and the other operand is in a working register called WR. Draw a diagram similar to Fig. 4.7 which shows all interconnections and clocking involving register AC and WR, and two circuit modules ALC1 and ALC2.

4.2 In Section 4.2 we have mentioned the use of tri-state output circuits for buffering. What special advantages do these circuits offer over conventional bi-state output circuits? Under what circumstances would the use of bi-state output circuits be quite adequate for buffering?

4.3 Discuss advantages/disadvantages of unidirectional common data bus vs bidirectional common data bus.

4.4 The processor generally contains PC, IR, AC, and SR registers. Most commercially available microprocessor units (MPUs) have a number of other internal working registers and 8,16 or more user-addressable registers. Give some examples of the need for these registers in user programs.

4.5 For each instruction a fetch followed by an interpret and execute cycle is required. However, interpretation and execution of the current instruction may cause conditions requiring interruption of the normal instruction sequencing operation to process these conditions and to decide whether the remaining program instructions should be accepted. Modify the flowchart of Fig. 4.14 to include this aspect.

4.6 In transferring data from memory we have included a WAIT state so that requested data is actually available in the MB register before transfer to processor is invoked. Under what conditions can this wait state be eliminated?

4.7 Through the use of different technologies and fabrication techniques, we have seen the emergence of higher and higher circuit densities on a single integrated circuit chip. We now have IC chips based on SSI, MSI, LSI, VLSI, etc., techniques. Give some examples of commercially available (or soon to be available) VLSI circuit chips and their applications.

4.8 Arrays of adder circuits can be used to build multiplication circuit modules. Draw the interconnection diagram of a 4-bit multiplication circuit module.

4.9 Draw a diagram for an 8-to-256 line decoder configuration using 4-to-16 line decoder IC chips.

4.10 Fig. 4.31 shows two common storage cells, one using magnetic core elements and the other constructed from multiple-emitter transistors. Give examples of some other storage cells and discuss their operation.

4.11 Large memory arrays require a large number of bits for addressing. Suggest a possible memory configuration where this requirement may be reduced.

REFERENCES

1. Blaaw, G. A. *Digital System Implementations*. Englewood Cliffs, (N.J.): Prentice-Hall, 1976.
2. Gschwind, H. W., and McCluskey, E. J. *Design of Digital Computers*. New York: Springer-Verlag, 1975.
3. Lancaster, D. *TTL Cookbook*. Indianapolis, (IN): Howard Sams & Co., 1974 ——— *CMOS Cookbook*. Indianapolis, (IN): Howard Sams & Co., 1977.
4. Morris, R. L., and Miller, J. R., eds. *Designing with TTL Integrated Circuits*. New York: McGraw-Hill, 1971.
5. Mowle, F. J. *A Systematic Approach to Digital Logic Design*. Reading, (MA): Addison-Wesley, 1976.
6. Osborne, A. *An Introduction to Microcomputers*, vols. I and II. Berkeley, (CA): Adam Osborne and Associates, 1976.

Chapter 5

FUNCTIONAL MODULES AND BASIC INTERCONNECTIONS

The evolution of a processing system is discussed in Chapter 4. In the same chapter, some of the essential components needed to build a computer system, like the arithmetic and logic circuits, memory circuits, decoder circuits, and the control and sequencing circuits, are also discussed. It has been indicated that most of the component circuits can be built from commercially available IC packages. The use of these IC packages to build a computer system greatly simplifies the overall design and subsequent interconnection of the various modules to form the system. The present chapter continues the subject treatment of the previous chapter. The main topics presented here are the organization and interconnection of the functional modules as well as the introduction of a very new and revolutionary class of significantly more complex IC packages, the microprocessors and their assortment of accompanying component circuits, which are the product of very highly developed LSI and VLSI (very large-scale integration) technology. The availability of these LSI packages is marked as a milestone in the flexible design and easy assembly of low-cost computer systems suitable for most of the small scale dedicated computational and/or control applications.

The reader is reminded that all of the functional modules, except the RTM modules (Section 5.4.1), will be investigated in much more detail in the next few chapters and so for all practical purposes can now be looked at as black boxes. He or she should mainly concentrate on understanding the flow of information and module interconnection techniques.

5.1 INTRODUCTION TO FUNCTIONAL MODULES

In Chapter 1, it has been indicated how a computer is invariably organized as an interconnection or network of physical modules. The advantages of this modular form of organization of a computer are many and include the following:

1. Each of the modules can be designed separately provided that they can subsequently be assembled into a working computer; this can significantly reduce the overall complexity of the design.
2. Each of the modules can be chosen to suit the operational requirements in terms of cost and performance effectiveness.
3. The amount of flexibility provided is very much enhanced; the modular design can be made in such a way that it is possible to start out with a system having a bare minimum of modules required to produce a working computer and to gradually add on additional modules as new requirements dictate or to replace existing modules with new ones.

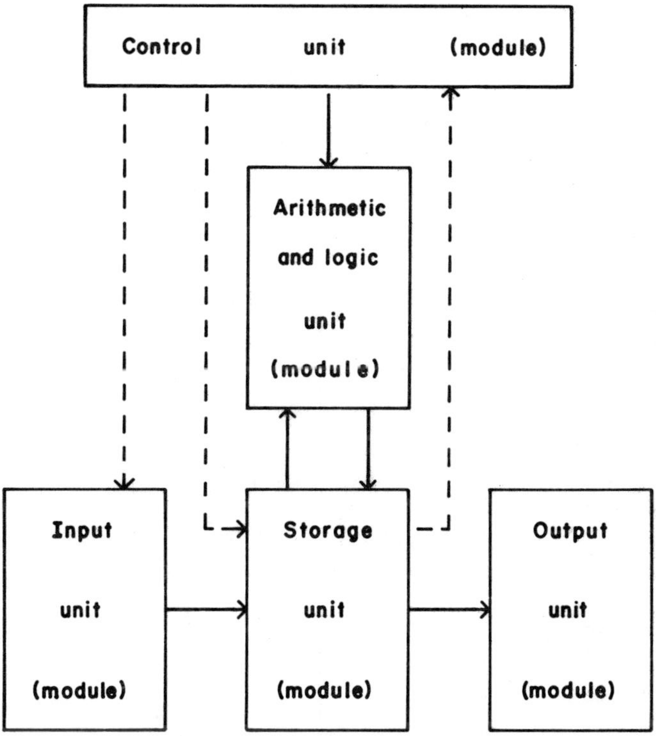

Figure 5.1 Modules of a digital computer. The dotted lines represent control signals, and the solid lines represent flows of information

The internal structures of the various modules constituting a computer and the form of the interconnection of the modules can vary widely, but the general configuration of a computer in modular block diagram form is that shown in Fig. 1.6 which is here redrawn in Fig. 5.1.

The block diagram in Fig. 5.1 shows the five major functional modules making up a computer. It is of course unnecessary to have a distinctly separate physical module to correspond to each of the five modules and in some cases, one functional module may actually be made up of a number of physically separate modules. For example, it is a usual practice for a physical I/O module to have the capability of performing both input and output.

The detailed structure of each of the five functional modules will be described in later chapters and for the moment our main interest is in understanding the overall organization and interconnection of the modules through consideration of the manner in which both data and control information are conveyed throughout the computer structure.

5.2 BUSING OF INFORMATION

Suppose we have four flip-flops each representing possibly one bit of each of four registers. Through the use of AND and OR logic gates, it is possible to control the transfer of information from one flip-flop to another by applying appropriate control signals. This is illustrated in Fig. 5.2. This

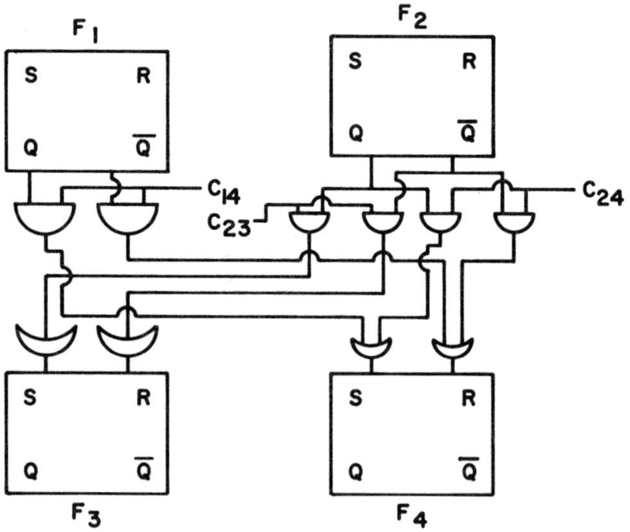

Figure 5.2 Gating of information among flip-flops

kind of controlled transferring of information among various storage elements is known as gating. Through application of a control signal at the terminal C_{14}, the content stored in flip-flop F_1 can be transferred orgated into flip-flop F_4 and similarly, application of a signal at the terminal C_{23} gates the content stored in flip-flop F_2 into flip-flop F_3, and application of a signal at terminal C_{24} gates the content stored in flip-flops F_2 into flip-flop F_4. It can be seen that if there is also a path between flip-flops F_1 and F_3, then a fourth set of gates needs to be added onto the gating circuit. This gating method can, of course, be expanded to perform controlled transfer of information among entire registers. Consistent with the algorithmic notation introduced in the previous chapter, we use the following notation to represent the presence of the necessary gating circuitry for a transfer of information from a register A to a register B

$$B \leftarrow A$$

If the transfer is effected for only a particular subset of bits in one or both of the registers, appropriate indices can be added to indicate the manner in which the transfer is to be done. For example, the expression

$$B(0{:}7) \leftarrow A$$

is used to represent the gating circuitry needed to transfer the contents of register A to the first 8 bits of register B. Register A is then taken to be 8 bits long. As another example, the following set of expressions:

$$B(0{:}1) \leftarrow A(0{:}1)$$
$$B(8{:}15) \leftarrow A(16{:}23)$$

is used to represent the presence of gating circuitry needed to transfer the contents stored in bits 0 and 1 of register A to bits 0 and 1 respectively of register B and those stored in bits 16 to 23 of register A to bits 8 to 15 of register B respectively.

The above gating method is adequate when the number of possible paths among the set of registers is very small. Since each possible path requires an individual subset of gating circuitry, the complexity of the entire gating circuitry becomes intolerable when the number of possible paths gets large. A generally adopted alternative is the method of busing of information mentioned briefly in Chapter 4. This busing method can actually be derived from the gating method by considering a transfer process as broken up into two steps: the transfer of information out of the source

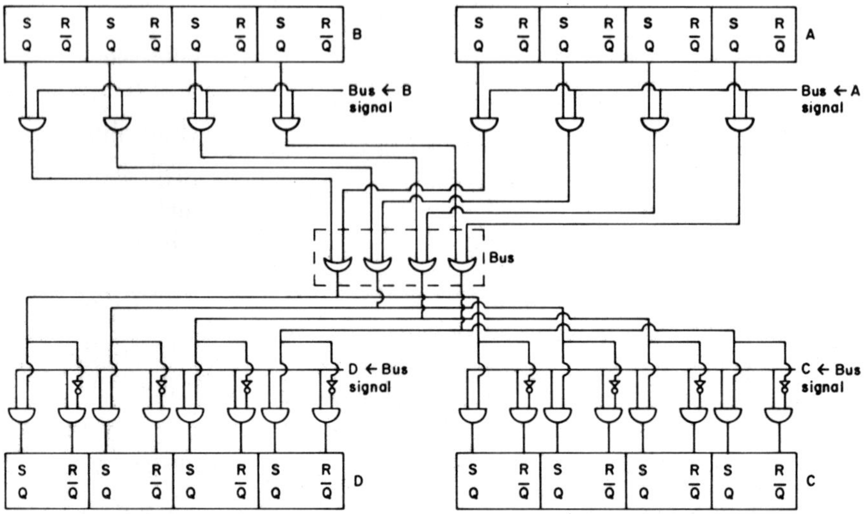

Figure 5.3 Busing circuitry

register into a common part of all the paths, known as the *bus*, and then the transfer of the information from the bus into the destination register. Thus, part of the circuitry, the bus, will be shared by all the transfer paths of concern, as a result of which the entire transfer circuitry will be greatly simplified. This busing method is illustrated in Fig. 5.3. The bus is shown enclosed in dotted lines.

It can be seen from Fig. 5.3 that a bus is essentially made up of a set of OR gates whose function is to accept information coming in from one of the registers having access to it and to transmit the information to one of the registers. Of course, an actual realization will most likely be made up of additional components such as drivers and sometimes even latches (e.g., IC 74125 introduced in Chapter 4). The sets of AND gates connecting the input and output terminals of the bus are used to select the particular pair of registers to be the source and destination registers.

It is convenient to use the diagrammatic form of representation shown in Fig. 5.4 to represent busing interconnection circuitry. The presence of a busing path from register R_1 to register R_2 can of course also be represented by the following two expressions

$$\text{Bus} \leftarrow R_1$$
$$R_2 \leftarrow \text{Bus}$$

The actual transfer is done by the simultaneous application of carefully synchronized control pulses to the corresponding terminals, or preferably by applying a level signal at the Bus $\leftarrow R_1$ terminal and a pulse signal at the $R_2 \leftarrow$ Bus terminal.

One obvious disadvantage of using a busing method for the transfer of information among registers is that only one of the paths sharing a common bus can be connected at one time. This problem can of course be partially alleviated by supplying a number of buses (or more precisely, subbuses) each being shared by a set of paths. For example, it is now common practice to provide a separate set of lines for transmitting address information and another for transmitting data information among various locations within a computer. Buses are usually bidirectional, being able to transfer information in either direction, sometimes simultaneously if sufficient lines are included.

There are now available commercially standard bus subsystems implemented on printed circuit boards that can be used for data transmission among functional modules in a computer system. They include both parallel and bit-serial buses. A parallel bus transfers all bits of information across different lines at the same time. An example of such a standard bus is the Altair-680B 6800 System Bus which has eight bidirectional lines for data transmission, sixteen unidirectional lines for address transmission and nine control lines for various control signal transmissions and is thus a parallel bus. Serial transmission of bits using bit-serial buses requires special attention and will be discussed in more detail in Chapter 8. These buses can also be used to connect the processor to external I/O devices through the use of interfaces. This topic will be considered in Chapter 8.

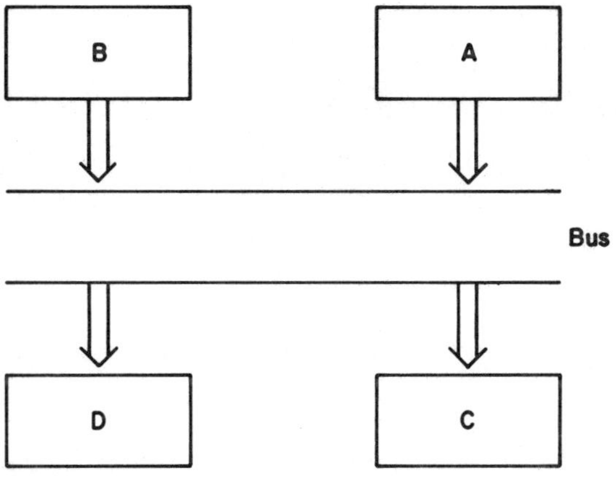

Figure 5.4 Busing representation: paths are identical to those in Fig. 5.3

5.3 ORGANIZATION AND INTERCONNECTION OF MODULES

Through the use of buses, functional modules can be conveniently interconnected to form computer systems. We first describe the organization of a very simple computer system, and then in a later section, examine certain hardware implications for various forms of interconnection of the modules.

Let us look at a very simple processing system obtained through the interconnection of a number of functional modules which are readily available as commercial integrated chips together with a number of other discrete components. A block diagram of the system is shown in Fig. 5.5. In the diagram, a total of eight boxes representing major components are interconnected through the use of buses and various control signal lines. We can of course allow various instruction and data widths, but we shall for simplicity limit our attention to instruction lengths of 8 bits and data lengths of 4 bits. Thus data paths that are used for transmission of instructions are assumed to be 8 bits wide and those for data transmissions are assumed to be 4 bits wide.

The major components making up the system are the following:

1. ALU—a simple 4-bit adder (e.g. 7483) or more sophisticated multiple-function arithmetic logic unit capable of directly performing addition, subtraction, and various other basic arithmetic and logical operations (e.g. 74181).
2. C—the carry-bit flip-flop.
3. OPA and OPB—these two operand registers can be implemented using any available 4-bit registers.
4. R_1-R_4, a collection of four 4-bit registers in which the operands, temporary results, and final results are stored. These can be implemented by a variety of RAM memory (e.g. 74170).
5. PC—the program counter register PC is a 4-bit counter (e.g. 74161).
6. MI—a 16 word by 8-bit RAM memory which is used to store instructions constituting a program to be executed in the processing system (e.g. 7489).
7. K—is the control module.

Some other components not explicitly indicated in Fig. 5.5 are the following:

1. A set of 4 switches for transferring data inputs into the system, and

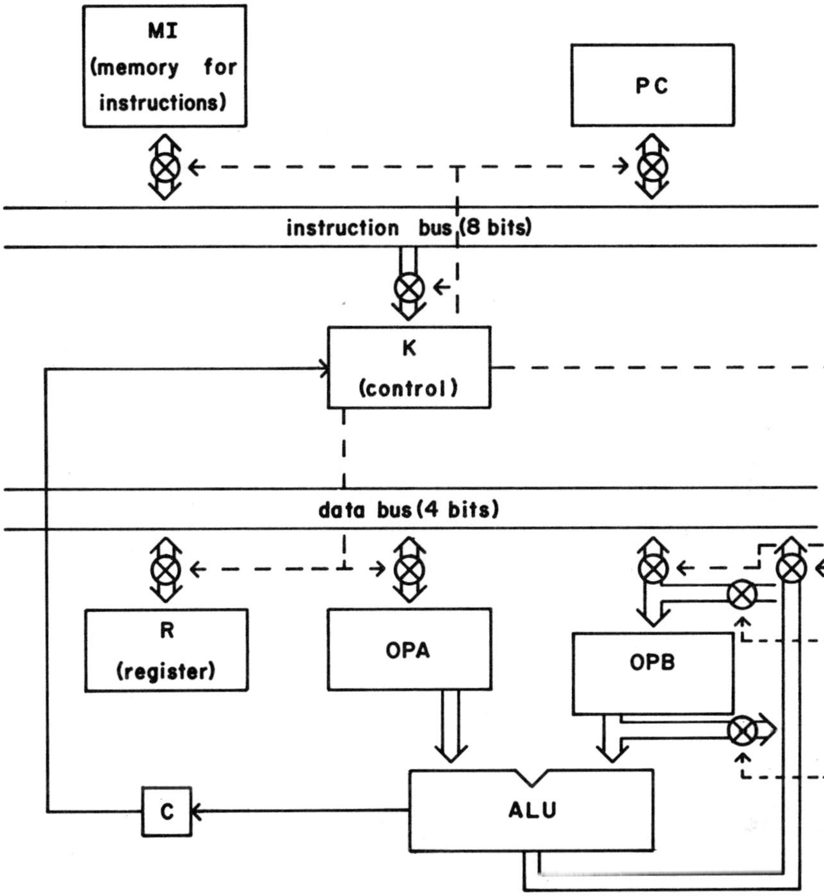

Figure 5.5 A simple computer. The dotted lines represent control signal lines

another set of display elements, e.g., lamps or LEDs (light emitting diodes) for data output purposes.

2. Facilities for manually loading instructions into any location in MI; this usually involves simply the incorporation of a set of switches for specifying the instruction and another push button for initiating the instruction loading.

Without consideration of the exact format of each instruction and data item at this point, let us see what the system can be expected to perform and what kind of control signals need to be generated by the control module K.

We can classify the instructions into four types as given in the following. At the beginning of an instruction cycle, the value in PC is used to select an instruction from MI to be executed. The value in PC is then incremented by one to point to the next instruction. The selected instruction is decoded in the control unit and appropriate control signals are generated:

1. If the instruction is a register-to-register instruction, e.g. an addition, incrementation or clearing, the operand(s) are first loaded from the appropriate register(s) in *R* (specified in the instruction) into OPA or/and OPB and then the required operation is performed. The result is then routed back to the destination register in R.
2. If the instruction is a branch instruction, the branching condition is tested first (in our simple system, this involves only testing the bit C) and depending on the result of the testing, the value in PC is changed to the target address or left unchanged.
3. If the instruction is an input instruction, the data is loaded into OPB and then transferred into the appropriate register in *R* through ALU. An output instruction operates in the reverse fashion: data in the designated register is loaded into OPB and then transferred out.
4. A stop or halt simply terminates execution.

The various control signals that are used to achieve the above operations are generated by the control unit K. It is not difficult to see that K is just a sequential circuit which is capable of generating a number of output signals used to synchronize and to effect the various gating and busing operations. The detailed discussion of the design of a control unit will be taken up in a later chapter, but it is at the present moment helpful to look at a few points concerning a possible design of the control unit K for our simple processing system.

The execution of an instruction can be divided into two parts: the *instruction fetch* and the *instruction execute*. During the instruction fetch, the content in PC is used to fetch the appropriate instruction from MI which is then stored in an instruction register in the control unit K and at the same time, the content of PC is incremented by 1 so that PC now points to the next instruction in MI. During the instruction execute, the instruction is decoded and then the required operations are performed.

As a result, the execution of an instruction requires two clock periods, one for the instruction fetch, and another for the instruction execute.

As an example, the instruction execute of a register-to-register instruction is achieved in three stages. During stage 1, one operand from one of the registers in *R* is loaded into OPA and during stage 2 the other operand, if there is any, is loaded from R into OPB. The specified operation and the subsequent storing into *R* of the result are then done in the third stage. To simplify the needed logic, it is convenient to use a 3-phase clock, that is, a clock whose output is made up of three distinct states per period. This can be accomplished with a ring counter (see Chapter 3) and is illustrated in Fig. 5.6. Note that one or more of the three phases may not be used for a branch, I/O or halt instruction.

Only four registers are available for data storage in the simple system. Thus only 4 bits are needed to completely specify the two operands involved in the register-to-register instructions. Also, because MI is limited to only 16 words, 4 bits are sufficient to specify a branching target address. This leaves us with 4 bits for specifying an operation code, for a total of 16 possible distinct operations. Thus the following is a set of possible instruction formats:

register-to-register

op code	1st operand	2nd operand
bit 0 … 3	4 … 5	6 … 7

branch instruction

op code	branch address
0 … 3	4 … 7

I/O instruction

op code	operand	not used
0 … 3	4 … 5	6 … 7

stop instruction

op code	not used
0 – 3	4 – 7

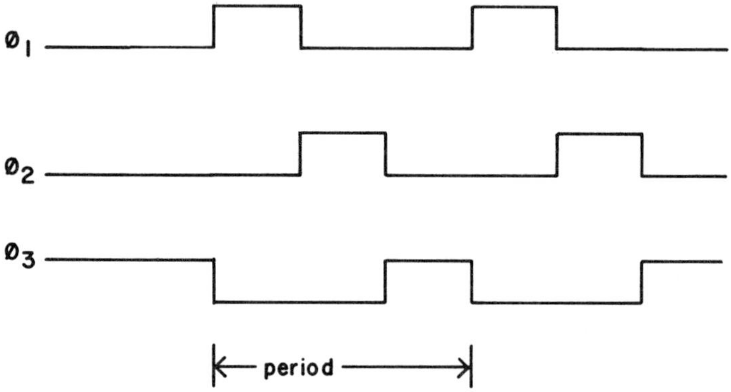

state 1:	$\phi_1 = 1$	$\phi_2 = 0$	$\phi_3 = 0$
state 2:	$\phi_1 = 0$	$\phi_2 = 1$	$\phi_3 = 0$
state 3:	$\phi_1 = 0$	$\phi_2 = 0$	$\phi_3 = 1$

Figure 5.6 3-Phase clock signals

The detailed design of the formats for both the instructions and data will be left as an exercise to the reader.

5.4 EXAMPLES OF FUNCTIONAL MODULES

In the last section, we have worked with relatively simple functional modules. Here we attempt to give some limited examples of certain more complex available functional modules that can be conveniently interconnected to form computer systems. If the readers find the material in this section confusing, they are advised to skip over to Chapters 6 and 7 and come back after they have finished those two chapters. Due to the enormous number of different computers of different scales, it is of course impossible to even identify representative systems. Our main objective here is to further solidify the concept of functional modules and their interconnections rather than to provide an exhaustive display of various structures of functional modules and the possible methods of their interconnections.

5.4.1 Register Transfer Modules

The concept of Register Transfer Level (RTL) was introduced by Bell, Grason, and Newell [1]. In this approach to the realization of a computer, the building blocks are Register Transfer Modules (RTM) which are a set of predesigned physical modules implementing logical functional modules ranging from simple registers to very complex control units.

We can normally classify the RT modules into 6 different types:

1. A transducer *T* for changing the represention of data between forms used external to the computer and forms used within the computer (the I/O modules).
2. A link L for transmitting information from one place to another in the computer (the bus subsystem).
3. A switch S for selectively gating information into and out of a link which is used for rerouting information.
4. A data operation D for transforming data or producing new information from existing information (the ALU).
5. A memory M for storing information (main memory or registers).
6. A control K for generating the control signals used in the computer (the control unit CU).

Within each type, a user is given a number of modules of varying characteristics to choose from.

When a processing system is to be designed, a flowchart describing the operation of the system is first drawn and then a set of RT modules is selected to configure the specified computer architecture around a system bus (better known as a universal bus). Generally, (close to) a one-to-one correspondence between the elements of logical organization and physical realization is established for the processor design.

An example of a general-purpose computer designed using RTMs is shown in Fig. 5.7. The logical organization shows a set of independent registers, such as Program Counter (PC), Stack Pointer (SP), Instruction Register (IR), Accumulator (ACC), Memory Address register (MAR), Memory Data Register (MDR), Lower and Upper Bound Register (LB,UB) for restricting user program references to within certain area of memory, six general registers (REG), two base registers (BASE), and

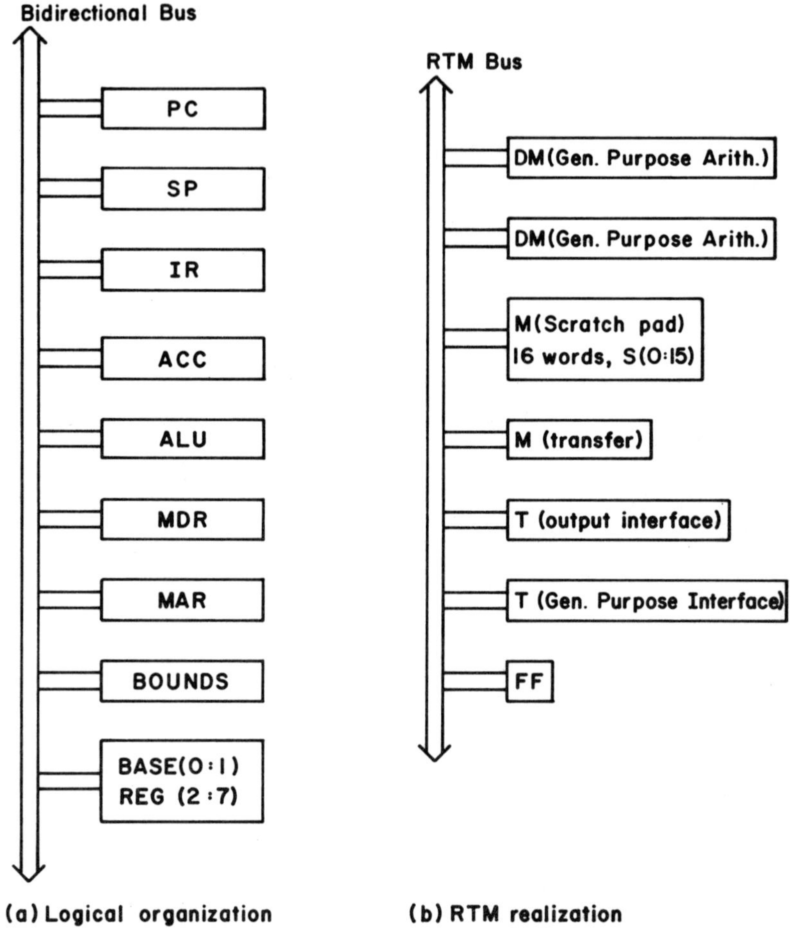

Figure 5.7 Logical organization and RTM realization of a processor

an arithmetic and logic unit (ALU). The functional roles of these registers will be described in more detail in Chapter 7. Transfers to and from these components (except MAR) take place via a bidirectional bus as shown. For simplicity, the control unit (to achieve sequencing of the transfers) is not shown.

Shown in Fig. 5.7(b) is an RTM realization of the logical system in Fig. 5.7(a). Since all the registers except IR and MDR are similar in characteristics, assuming a word size of 16-bits, an RTM scratch pad memory of 16 words can meet the requirements adequately as shown. A different choice is made for IR, namely, M(transfer), because in this module, each bit is available as a Boolean output, and the upper as well as lower half of this module can be written or read directly as two separate fields f_1 and f_2. For MDR, the T(general purpose interface) module is chosen to provide direct bidirectional transfers. For the ALU, two DM (general purpose arithmetic unit) modules are chosen. Each DM module contains two working registers to receive data, perform the desired operation, and output the result. A flip-flop, FF, is used for the link portion of ACC. For sequencing of transfers, a set of evoke modules, K, may be used to establish hardwired control. However, to allow for reasonably complex and modifiable sequencing operations, we can select a Microprogrammed Controller, K(MPC) (See Chapter 7 for a description on microprogrammed control). For a more detailed discussion on the various RTM modules, see Reference 1.

The above brief discussion of RTM demonstrates the amount of freedom with which the processor of a specified architecture can be con-

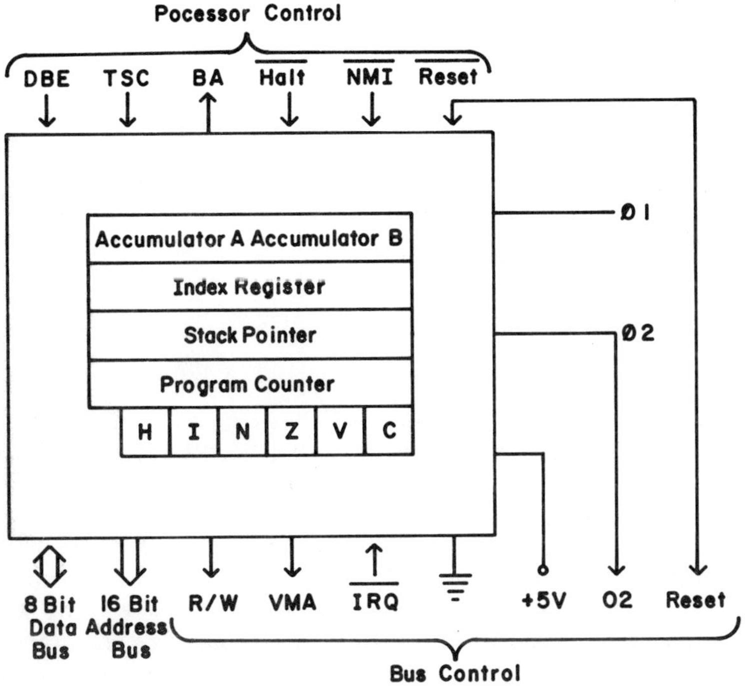

Figure 5.8 MC6800 Microprocessor and input/output lines

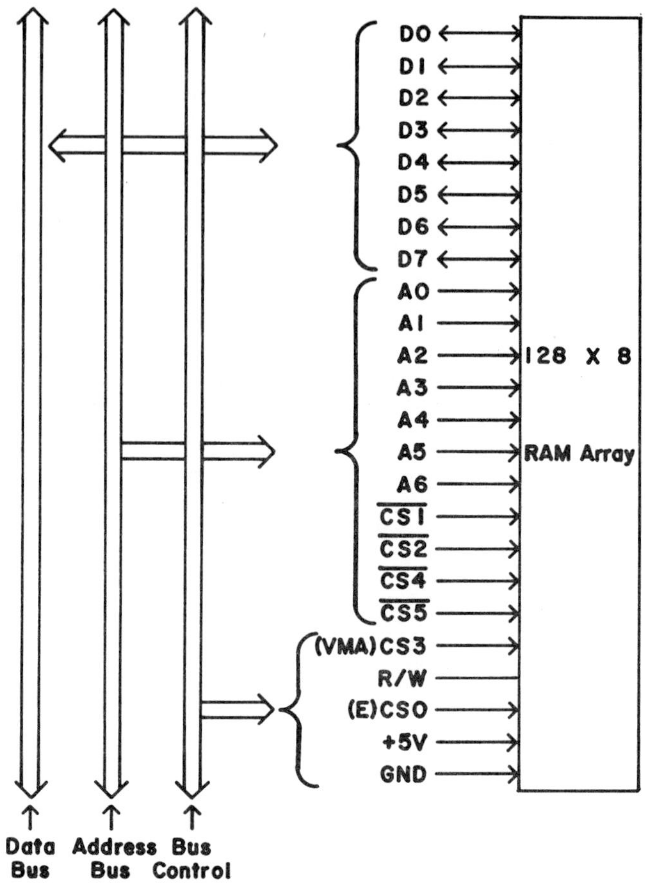

Figure 5.9 MCM6810A RAM BUS INTERFACE

structed using RTM modules, a fact which is especially important in the design of special-purpose systems. This high degree of freedom is not available in a microprocessor-based system, which we are now going to introduce. However, in the area of interconnection of the processor to the rest of the system, we will see that there is more freedom and flexibility in a microprocessor-based system than in the RTM-based system. In addition, a microprocessor-based system with a reasonable computing capability can be constructed at a cost of a few hundred dollars while a comparable RTM-based system can cost up to a few thousand dollars.

5.4.2 Introduction to Microprocessor Systems

The present rapid advent of integrated circuit technology has made possible the design of computer systems in which the building blocks are much more functionally complex than the RTMs introduced in the previous section. In fact, most of the functional modules shown in Fig. 5.1 are now widely available as very low cost single-integrated circuit chips in dual-inline-packages (DIPs). These DIPs can then be interconnected to form computer systems. The interconnections are usually done on accompanying logic breadboards which contain among other things socket strips for plugging in DIPs and sometimes even power supplies for easy

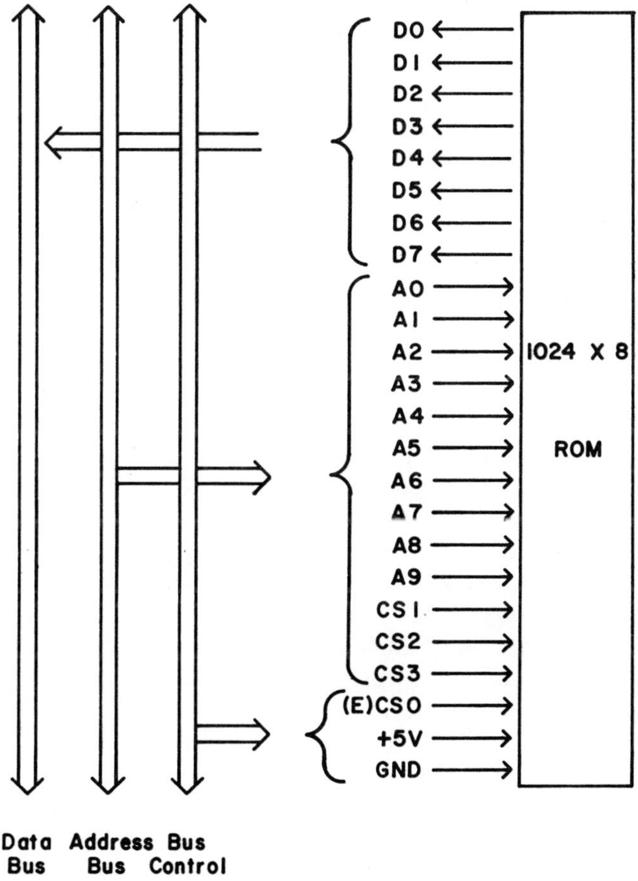

Figure 5.10 MCM6830A ROM BUS INTERFACE

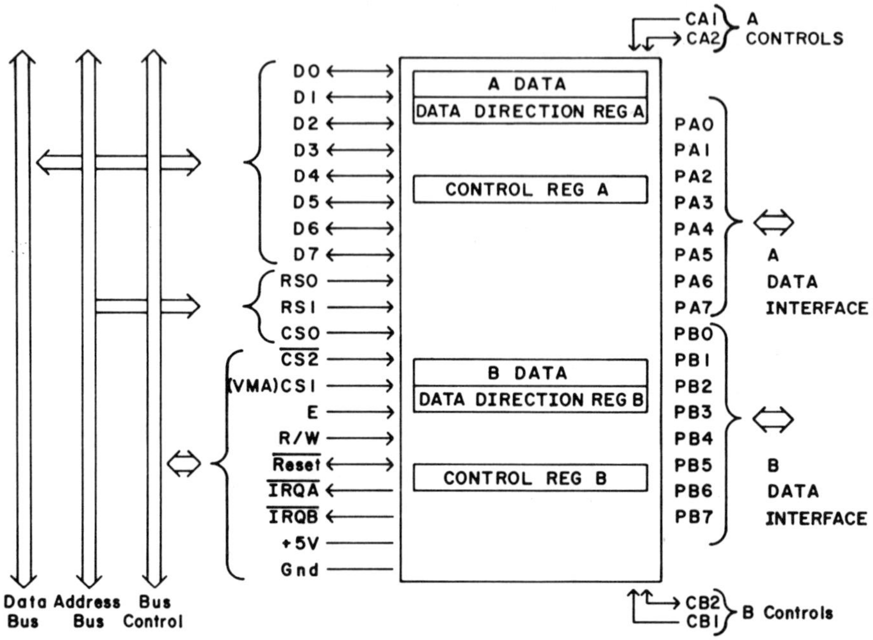

Figure 5.11 MC6820 PIA BUS INTERFACE

use. The CPU is generally known as a *microprocessor* (MPU) and a system based on a microprocessor is generally known as a *microcomputer*. It is sometimes argued that the distinction between a microcomputer and a mini-computer should be based on the areas of application and overall costs of the systems, microcomputers being low-cost devices used mainly in small-scale special purpose applications, like pocket calculators, small intelligent terminals, and oscilloscopes with some arithmetic and logic capabilities.

The Motorola MC6800 family of parts has been designed to set the standard for microcomputer system architectures. To build a minimum system requires an MC6800MPU, an MCM6830A ROM, an MCM6810A RAM and an MC6820 PIA (peripheral interface adapter). Block diagrams of these 4 modules are shown in Figs. 5.8–5.11.

The block diagram for the MC6800 MPU shows some of the important registers residing in the MPU and four sets of input and output lines which are used to interface the MPU to other components in the system. A set of 8 lines is needed to interface to the data bus for transmission and reception of data, and another set of 16 lines is needed to transmit the

16-bit address to either RAM,ROM or input/output devices. The signal lines R/W, vma, $\overline{\text{IRQ}}$, +5V, ϕ_1, ϕ_2 and $\overline{\text{Reset}}$ constitute the set of bus control lines. The fourth set of lines is the set of processor control lines DBE, TSC, BA, $\overline{\text{Halt}}$, $\overline{\text{NMI}}$, and $\overline{\text{Reset}}$ used to control certain aspects of the operation of the MPU.

Block diagrams for the other 3 modules are shown in Figs. 5.9–5.11. The MCM6810A RAM module has 128 bytes of static random access memory (see Chapters 4 and 6 on the discussion of various types of memory). Chip select (CS) signal lines are available for addressing a particular module in case more than one memory module is present in the system; decoding of the CS signals enables the appropriate chip module and disables the rest. The MCM6830A ROM module is organized as 1024 bytes of static read-only-memory and has four chip select lines. More detailed block diagrams of these two types of modules can be found in Chapter 6. The MC-6820 PIA provides the universal means of interfacing peripheral devices to the MPU through two 8-bit bidirectional data interface lines (PA and PB lines) and four control lines (CA and CB lines). A more detailed description of this module can be found in Chapter 8. The formation of a computer system comes down essentially to the interconnection, sometimes involving additional buffers, of the three sets of interfacing lines from each of the modules: the data bus, the address bus, and the bus control. Thus the design of a microprocessor-based system is a comparatively less difficult task and can be undertaken by persons with a background in logic design but only a limited amount of knowledge in electronics.

We describe in the following paragraphs some microcomputer systems based on a few of the more popular microprocessors. It should be noted that for some microcomputer systems, specific components have to be used while for some other systems, certain limited choices may be available for some components. Manufacturer's design manuals have to be studied carefully when microcomputer systems are developed.

An Intel-8080A MPU based computer system is shown in Fig. 5.12 and a Motorola-M6800 based computer system is shown in Fig. 5.13. Both systems are designed around a system bus with various components connected into complex configurations. Both the 8080A and M6800 have a 16-bit address bus, an 8-bit bidirectional data bus and fully compatible T^2L control outputs. Besides supporting up to 64 K bytes of RAM, they can address a large number of peripheral devices, providing for practically unlimited system expansion. Peripheral interface circuits receive their inputs on the control and address buses and return their data outputs onto the data bus.

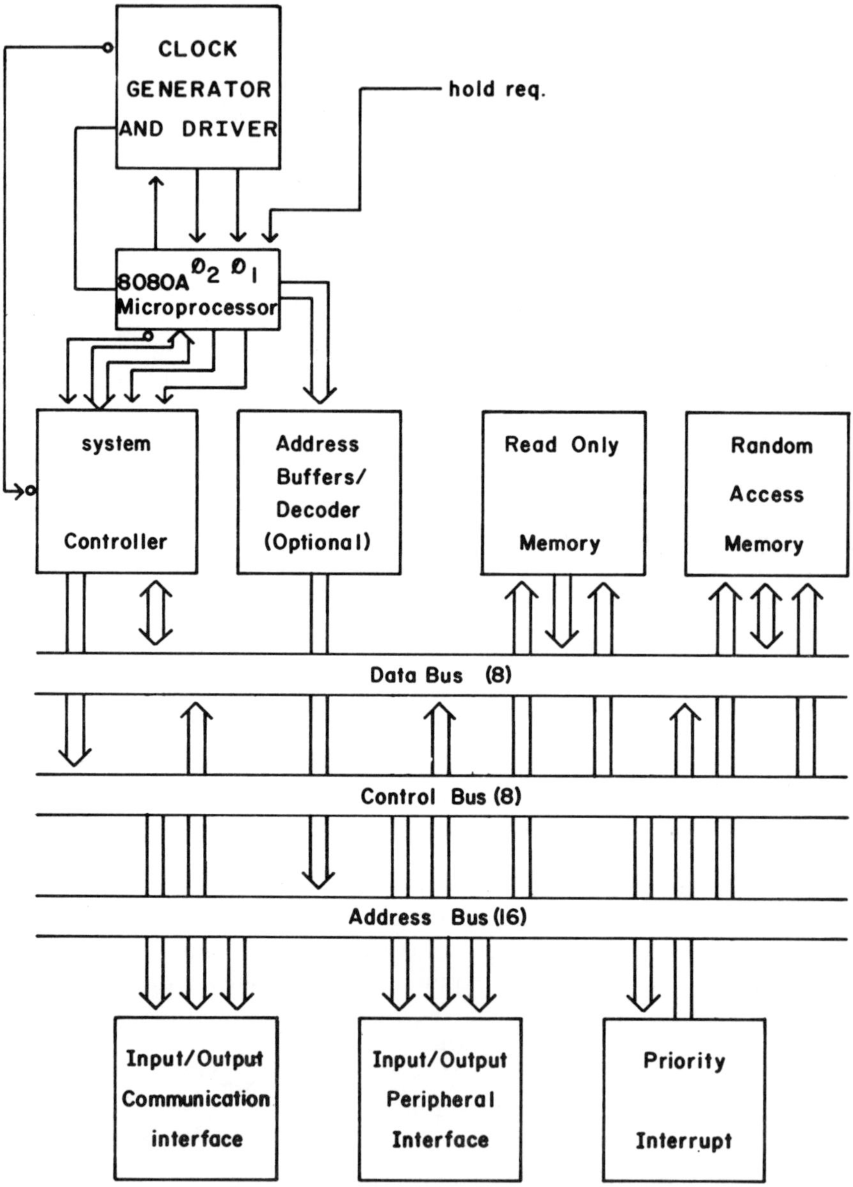

Figure 5.12 An Intel-8080A based microcomputer system

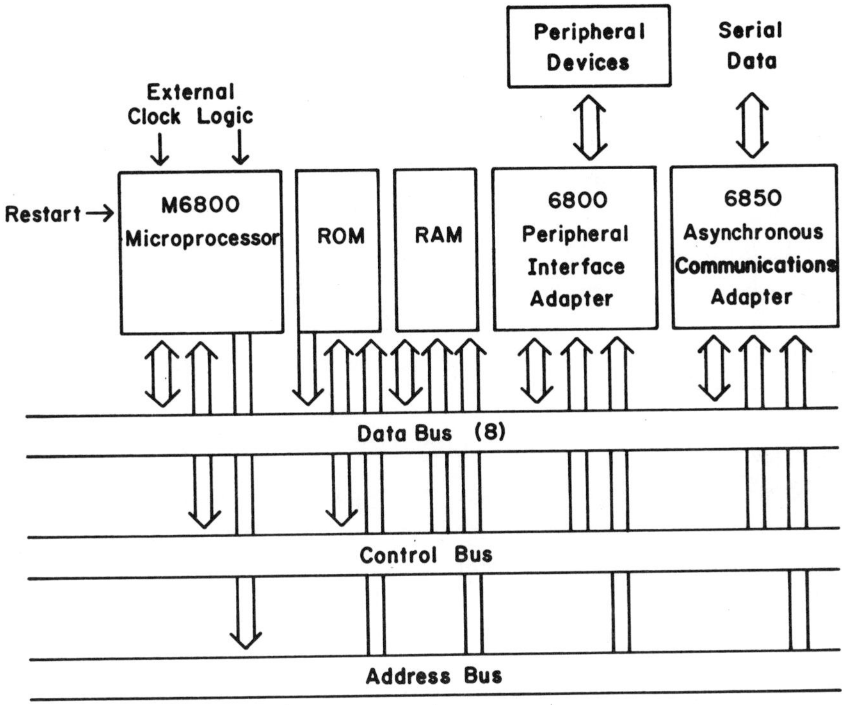

Figure 5.13 Motorola-M6800 based microcomputer system

A Fairchild-3850 based F8 computer system using only two major components is shown in Fig. 5.14. The two components are the 3850 CPU and a program storage unit, the 3851 PSU. Although we shall defer the description of the internal structures of the components to later chapters, it is worth noting here that the F8 has a number of distinctive features:

1. A small amount of RAM is implemented within the CPU to be used as scratch pad memory which is not directly accessible to the programmer.
2. Memory addressing logic is implemented in the memory devices in the above specific F8 configuration, the 3851 PSU, and not in the CPU.
3. I/O ports are implemented in the CPU and memory devices, rather than as separate I/O devices.

In all cases, some additional compatible modules can readily be added

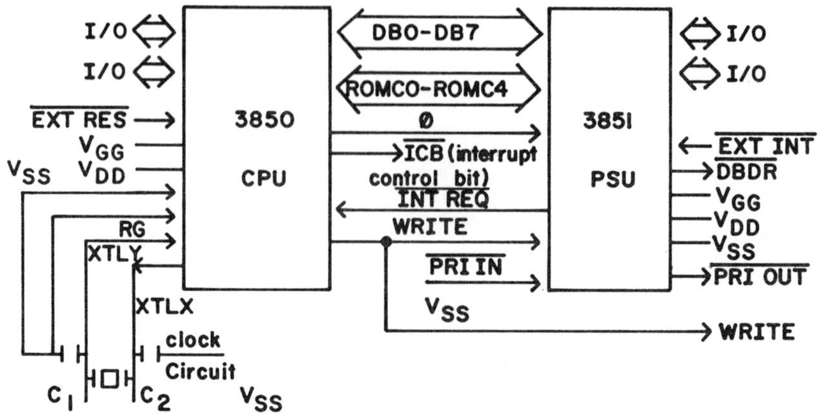

Figure 5.14 A basic, two-device F8 configuration

on to the existing system to increase its capability. For example, the basic 2-device F8 system shown in Fig. 5.14 can easily be expanded into a 3-device system through the use of an additional program storage unit (PSU) resulting in the system shown in Fig. 5.15. In fact, a maximum of 64 of the 3851 PSUs can be connected to a 3850 MPU in this way.

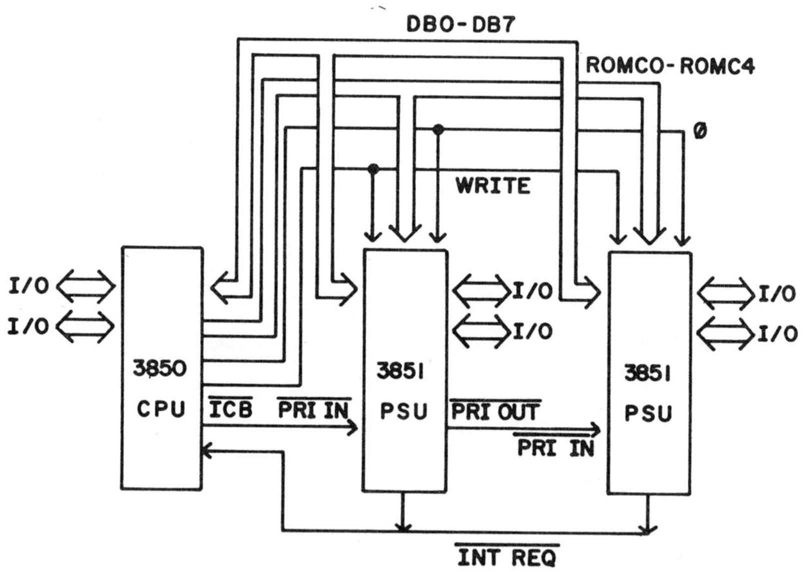

Figure 5.15 A 3-device F8 system

EXERCISES

5.1 The symbols "$\uparrow$" and "$\downarrow$" are used to indicate left and right rotation respectively. Let AR be a 4-bit register. Construct a detailed logic diagram of the hardware needed to accomplish the following transfer operation, rotating the content in AR left one place:

$$AR \leftarrow \uparrow AR$$

Use master-slave flip-flops for AR.

5.2 Let MA and DC be 4-bit registers and MB and AC be 2-bit registers. The following set of transfer operations are required:

$$MA \leftarrow DC \qquad DC(0{:}1) \leftarrow MB$$
$$AC \leftarrow MB \qquad DC(2{:}3) \leftarrow MB$$
$$MB \leftarrow AC$$

Construct a busing system similar to that described in Section 5.2 needed to accomplish the above set of transfers. Assume all the registers are of master-slave flip-flops.

5.3 Repeat Problem 5.2 using only one single bus which is 2 bits wide in the busing system.

5.4 Complete the design of the instruction and data formats for the simple computer system described in Section 5.3.

REFERENCES

1. Bell, C. G., Grason, J., and Newell, A. *Designing Computers and Digital Systems*. Maynard, MA: Digital Equipment Corporation, Digital Press, 1972.
2. Fairchild Corporation. *F8 User's Guide,* 1976.
3. Hilburn, J. L., and Julich, P. M. *Microcomputers/microprocessors: Hardware, Software and Applications*. Englewood Cliffs, N.J.: Prentice Hall, 1976.
4. Hill, F. J., and Peterson, G. R. *Digital Systems: Hardware Organization and Design*. New York: John Wiley & Sons, 1973.
5. Lesea, A., and Zaks, R. *Microprocessor Interfacing Techniques*. Berkley, CA: Sybex, 1977.
6. Motorola, Inc. *M6800 Microcomputer System Design Data,* 1976.
7. Wakerly, J. *Logic Design Projects Using Standard Integrated Circuits*. New York: John Wiley & Sons, 1976.

Chapter 6

MEMORY MODULES AND DEVICES

A memory module consists of storage elements, an address register connected to decoder and driver circuitry, a buffer register for reading information from the selected storage elements as well as for writing information into the storage elements, and a control line to enable the desired read or write operation. Depending on the nature of the storage elements it may also require refresh circuitry to periodically restore the contents of storage elements to desired levels. Also, in some cases the locations of selected storage elements may have to be moved to a read/write station before the actual read/write operation can be performed.

The variations in the characteristics of storage elements and memory modules stem from the need to satisfy a number of goals. The ideal memory module should provide a large storage capacity (billions or trillions or more bits of information), should provide rapid access for read/write operations (microseconds, nanoseconds, or less) and should be quite inexpensive (.01 cents or less per bit). Unfortunately, at current technological levels all these goals cannot be met at the same time. Therefore, we will discuss various memory modules which come close to satisfying one or more of these goals. A variety of these modules may be selected and integrated into a number of levels to provide some high-speed but relatively expensive random access primary memory capability and the rest as secondary or higher levels of larger capacity but less expensive and slower storage. The information processing requirements can then be met by moving information to primary memory for access by the processor and moving it to secondary levels when it is not required by the processor so as to make room for information from the higher levels which is required by the processor immediately. This arrangement is known as memory hierarchy. Consider the following two level hierarchical system:

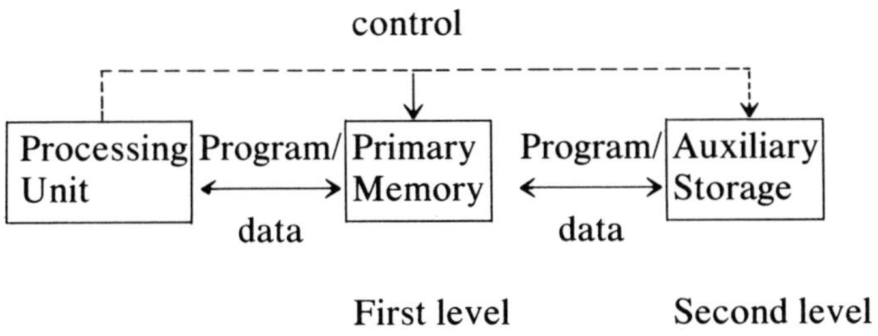

During execution of a program resident in primary memory the processing unit may make reference to data files or a segment of the program not resident in memory. The control signal from the processing unit will then cause the required information to be moved from auxiliary storage to primary memory. Since the auxiliary storage is slower than primary memory, the execution of the program must be delayed. If the mismatch between the speeds of operation at the two levels of storage, is very high, e.g., magnetic-core primary memory with access time of 1 μsec. and magnetic disc secondary storage with access time of 100,000 μsec. then this delay can be very significant and may adversely affect the system performance. This performance can be improved considerably by introducing additional level(s) of hierarchy between the two types mentioned above. For example, in a three-level memory hierarchy the middle level may consist of storage elements with access times in the range of 100–1000 μsec. and 1/10th the cost of primary memory. We will discuss memory modules constructed from these elements in section 6.8. The memory hierarchy could also include a small high-speed memory with access time lower than .1 μsec. (bipolar transistor elements) at level 0, i.e., between the processing unit and primary memory. This memory could retain most frequently accessed data thus improving the overall system performance even further.

In this chapter, first we discuss some examples of memory technologies which are used for primary storage to be followed by examples for higher levels and finally some examples of special purpose memory modules.

6.1 LINEAR SELECT MAGNETIC CORE MEMORY

In a magnetic core memory, the elements are arranged on a single plane (2-dimensional organization). The elements in a given row represent one word of storage and the total number of elements in a given column represents the storage capacity in words, as shown in Fig. 6.1.

Only two wires are threaded through each core. For a read operation,

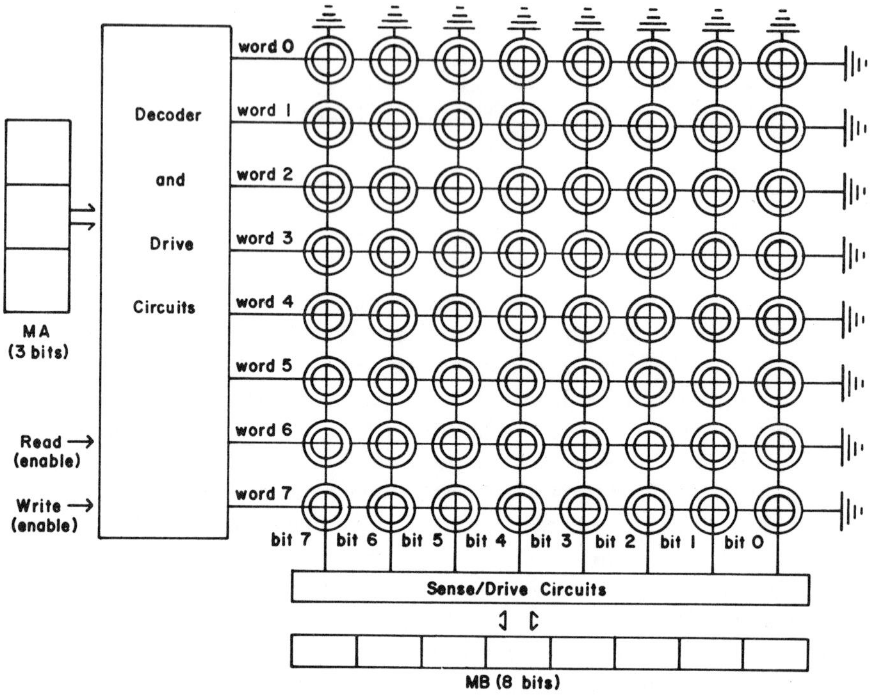

Figure 6.1 Organization of an 8-word, 8-bit magnetic core memory module with linear selection

the word line corresponding to the address contained in the MA register is energized with a current level which is high enough to switch the state of the cores. This current has an appropriate direction (right to left in the figure shown) so that all cores containing one will go through a flux change (because of switching from clockwise to counter-clockwise direction of magnetization). The cores going through this flux change will induce a current on the sense wire which, after proper amplification, will set the corresponding bits in the MB register. The read operation performs correctly but it also clears (writes a zero) in the magnetic core elements of the selected word during this operation. Thus, a read operation must be followed by a writing of the contents of MB back into the selected word if it is desired to restore the previously stored information. In some applications, such as buffering operations, this write-back may not be necessary.

For writing information from MB into the selected word, first a zero is written into all bits of this word. Note that this is not necessary if read and write operations are performed alternately because every read (without restoration) automatically clears (writes a zero) in all core elements of the selected word. For a given bit pattern in the MB register, the drive circuits are energized which causes the current to flow in the bit lines corresponding to every bit which has a value of 1. However, these currents have a level of only half of what is required to switch the states of the cores. Simultaneously, the selected word line carries drive current (from left to right) which is also half of the required level. This produces a collective (coincident) level of the full current required through the cores where a 1 is to be written, causing them to magnetize in the desired clockwise directon.

In ordinary applications both read and write operations will require two cycles of magnetization. However, in buffering operations between devices where information stored need only be read once and replaced by new incoming information, the speed is doubled. Since only two wires are threaded through each core, it is possible to reduce the size of the cores. Smaller cores can be switched at a faster rate (duration of magnetization cycle reduced) enabling a higher speed of operation. Thus, the linear-select memory organization is well suited for high speed applications where the size of the required memory capacity is small. However, this organization is not suitable for larger memories because of the rapid increase in the cost of the decoder and drive circuits.

6.2 COINCIDENT CURRENT MAGNETIC CORE MEMORY

In Section 4.5.1, we described a magnetic core memory cell based on coincident current selection. Using this cell, memory modules of any desired capacity can be configured. The read/write operations are usually assumed to access one full word, i.e., all bits of the word at the specified address. It is, therefore, natural to conceive of such memory modules as a stack of planes where each plane contains as many bits as the number of words in memory. The organization of a single memory plane for magnetic-core memory using the coincident-current technique is shown in Fig. 6.2. The magnetic cores range in diameter from .01'' to .03''. A large number of these cores can be organized in a single plane, e.g., 1024 or 4096 cores per plane. The limit on the size of any one plane is set by the signal (from selected core)-to-noise (from other cores carrying half select current) ratio which is tolerable on the output signal on sense wire. Another important parameter to consider in computer memory is the access time or the length of time it takes to read information from memory (including the time to restore the contents of the word read if readout is

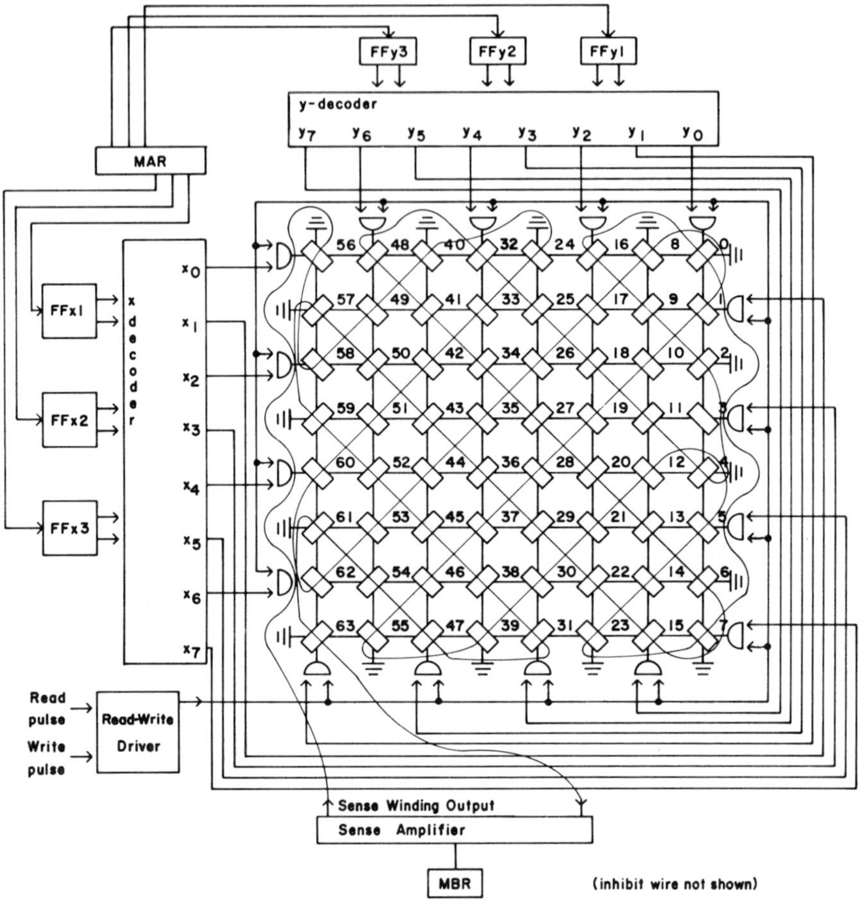

Figure 6.2 64-bit core memory plane. The wiring of x, y and sense wires has been organized in a manner that while the selected core receives full switching current, others receiving ½ selection current do not have any net effect on the sense wire. Of the total 6 bits in the MAR register, the right three bits are used for the x-decoder and the left three bits are used for the y-decoder. The contents of MAR, e.g. 0 1 0 1 1 0 will select x_6 and y_2 lines which is the address of core number 22. The 64 cores are addressed from 0 to 63 i.e. memory of 64 words with one bit per word. To have m-bits per word a stack of m identical core planes will be constructed and wired so that all bits (cores) of the addressed word are selected simultaneously. There will be one sense amplifier per plane leading to the corresponding bit of the MBR. The inhibit wire which has not been shown in the figure runs parallel to x-wires but through all the cores of the same plane.

destructive) or write information into memory. The reading or writing of information into cores requires two consecutive pulses. In a read operation, the first pulse is used to write a zero in all the cores of the selected word while the sense amplifier is on. This loads the MBR with the contents of the memory in the selected word. The second pulse is used to restore the contents of the cores typically by using the MBR output but keeping the sense amplifier disabled; each pulse may be about .4 μsec in duration with about a .2 μsec delay between pulses, thus requiring a total of 1 μsec. The same is true for a write operation. We can use a switching pulse of a shorter duration for suitably designed cores giving a faster access time in the range of 300-500 nsec. The characteristics of magnetic materials make it difficult to make significant improvements.

As shown in Fig. 6.2, a typical design using magnetic cores requires decoder circuits, drivers, sense amplifiers, inhibit amplifiers, MAR, and MBR. The driver circuits are required because current levels of about 100 mA are required on each of the select (and inhibit) lines. The address in MAR is externally decoded to select appropriate X, Y lines. The am-

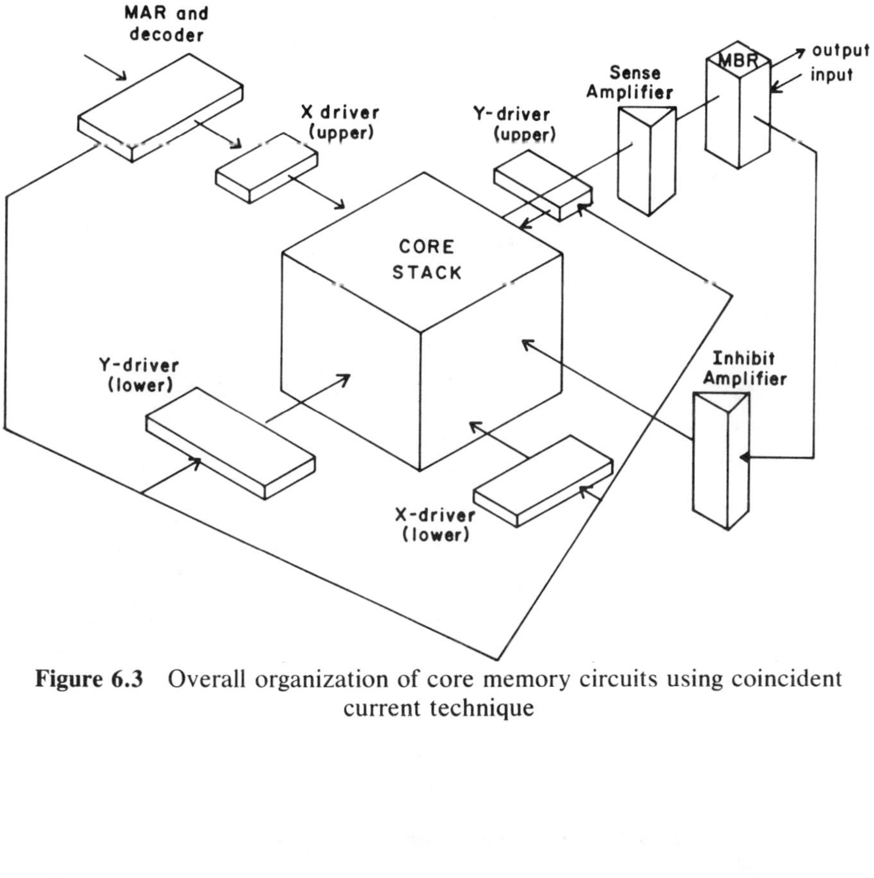

Figure 6.3 Overall organization of core memory circuits using coincident current technique

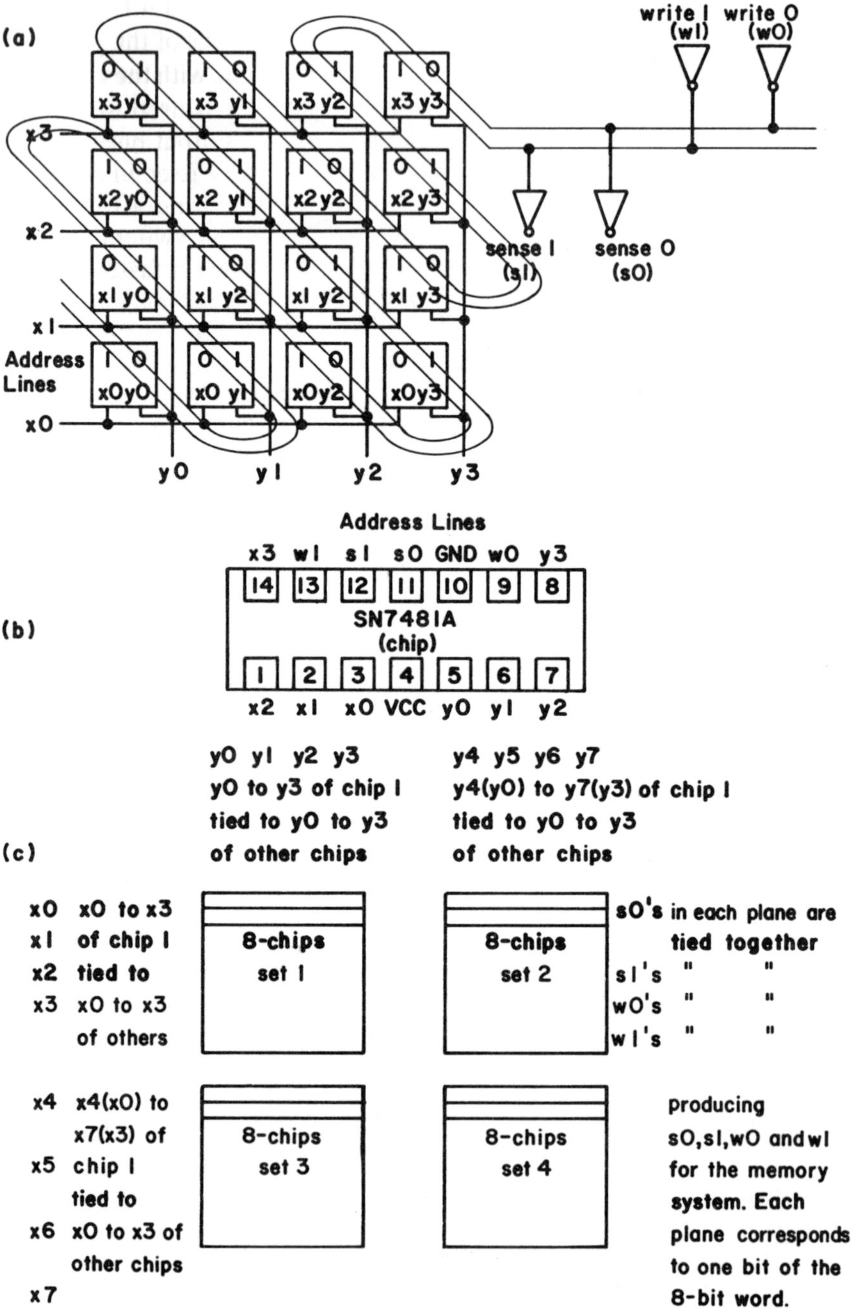

Figure 6.4 Semiconductor memory array
(a) Logical organization of a 16-bit memory
(b) Memory array chip
(c) 64 8-bit word memory using the arrays shown in (b)

plifier is needed to amplify induced signals on the sense lines and generate corresponding one or zero signal levels required to set/reset MBR flip-flops. The overall organization of coincident current magnetic core memories is shown in Fig. 6.3.

6.3 COINCIDENT CURRENT SEMICONDUCTOR MEMORY

In Section 4.5.1, we described a semiconductor storage cell using coincident current selection. Using this cell, memory modules can be constructed in much the same manner as discussed above for magnetic core storage cells. However, there are a number of considerations which make these arrays quite different from core memory arrays. Firstly, the variation in speed (access time) is achieved by an appropriate choice of semiconductor technology, e.g., conventional bipolar transistors (TTL), MOS(FET), etc. Secondly, since decoder circuits and memory cells can use the same technology, the decoding function can be made internal to memory arrays, i.e., a memory array chip also includes a decoder circuitry. Also driver and read/write circuits are included in a memory chip. This is particularly appealing when the array size is in the range of a few hundred to a few thousand bits. Fig. 6.4 shows a 16-bit semiconductor memory. This is high speed memory using a TTL array of 16 flip-flops as discussed earlier and includes write (drive) and sense amplifiers for single selected bit write/readout. These memory arrays have access times of a few nanoseconds and are usually employed in high speed scratch-pad memory applications. For slower (cheaper) memories integrated circuit chips are organized in linear-select mode (word addressing rather than bit addressing) because it permits a high bit density in the fabrication of chips.

6.4 READ-ONLY-MEMORY (ROM) MODULES

Memories can be thought of as devices which perform translation or conversion of an arbitrary input-bit pattern (address bits) into some output-bit pattern (output from the addressed location). The correspondence between these inputs and outputs can be changed at will in the case of read/write memories, by writing new information into selected memory locations. In many instances we may wish to establish a fixed correspondence between inputs and outputs. For example, in character generators for displays we may provide 6-bit ASCII codes for the characters and get the patterns for the array of dots to be displayed on the screen. A given 6-bit input pattern will always generate the same output pattern.

This is a specific instance of code conversion. Trigonometric function generators also fall into this category. Another example of fixed input-output correspondence is process controllers. The same applies to a variety of programs which are frequently used in computers such as monitors containing routines for I/O devices, language interpreters, assemblers, or even compilers. In all of the above instances a memory without the usual write capability is not only adequate but desirable provided the information written once can be retained indefinitely even when the power is turned off or remains unaffected by system malfunctions. This type of memory is called read-only-memory (ROM). Initially, ROMs were employed in computer systems to store microprograms as a replacement for hardware logic in instruction decoding and execution (see Chapter 7). With the advent of microcomputers, they have become very popular for storing and retaining all kinds of machine programs needed frequently. The basic structure of read-only-memory is shown in Fig. 6.5. The true-false (1-0) correspondence between input and output lines is established by the presence or absence of coupling elements. The kind of coupling element determines the type of memory. For example, punched cards are

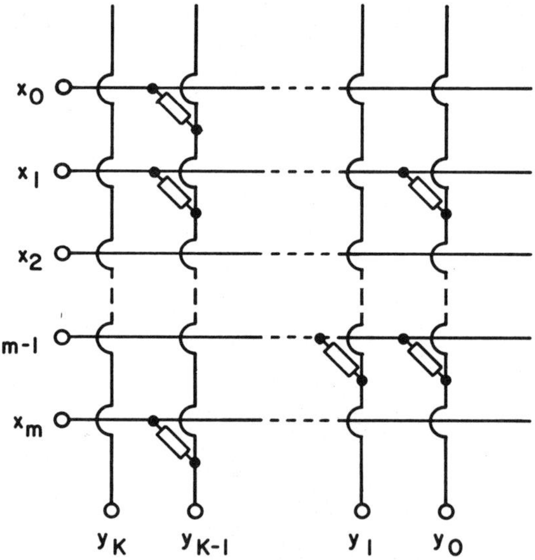

Figure 6.5 Structure of read-only-memory (ROM). There are m input or address lines and K output or data lines. Coupling elements appear between desired input output lines. In the illustration, whenever input at x_0, x_1 or x_m is true, the output y_{K-1} is true. Similarly whenever x_1 or x_{m-1} is true, y_0 is true and whenever x_{m-1} is true y_1 is true

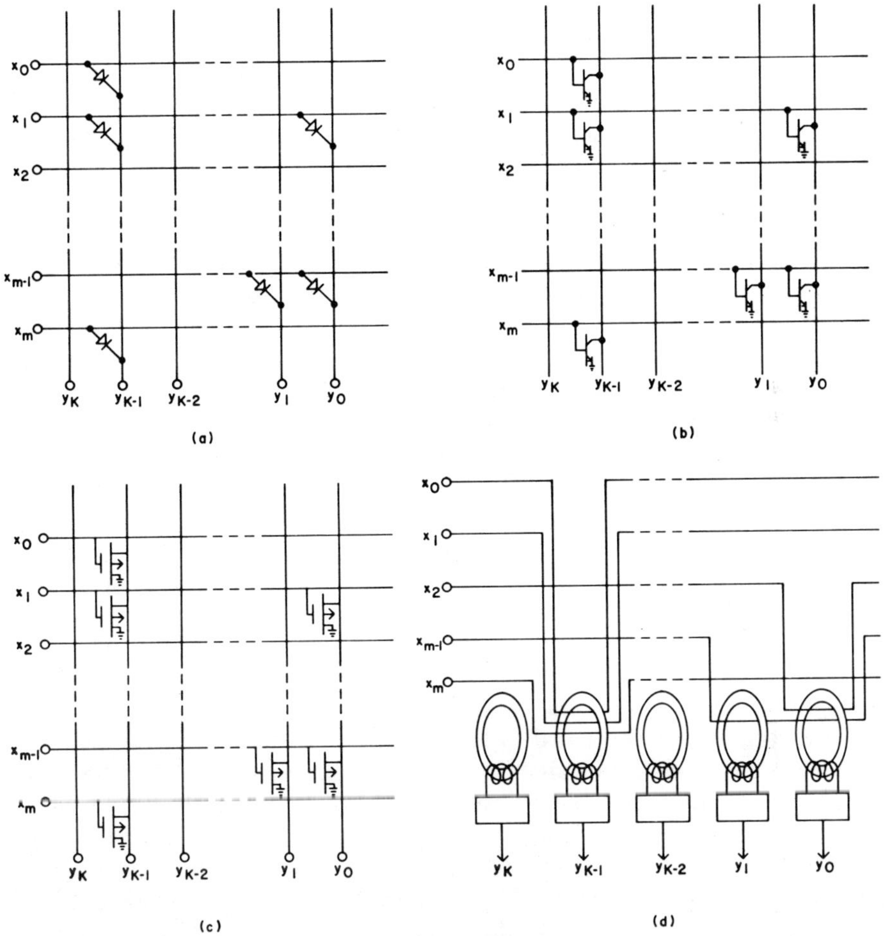

Figure 6.6 ROM arrays. During fabrication all possible coupling elements are prepared or created. Actual memory contents are created by masking, burning or removing contacts of those coupling elements which are not required

(a) Diode ROM array
(b) Bipolar ROM array
(c) MOSFET ROM array
(d) Transformer (inductor) ROM array

used to retain information wherein a hole indicates the presence of a coupling element in a specified position. However, most read-only-memories use passive or active electrical components as coupling elements. Examples of active elements are diodes, and bipolar and MOS semiconductor transistors; whereas resistors, capacitors, and inductors (transformers) are passive elements. All of these are used but they differ in terms of operating characteristics and cost. Also some fabrication techniques (not mentioned above) permit erasing of the entire memory contents and recreating new configurations of coupling element contacts. Examples of some physical realizations of ROM arrays are shown in Fig. 6.6. Popular ROM memory arrays come in word sizes of 4 or 8 bits and capacities of 256 to 1024 words customized to user supplied storage contents by the manufacturers. Field alterable and field programmable ROMs (PROMs) are also available but require special equipment for altering the contents or programming of these ROMs.

6.5 ASSOCIATIVE AND CONTENT-ADDRESSED MEMORY MODULE

In Sections 6.2–6.4, we have discussed the design and organization of random-access memory. Given the address (location) of any memory word, we could retrieve the contents of that memory location by activating the read control signal. Similarly, we could write into any location. This form of organization is known as location-addressed memory. There are many instances where we may want to access a word by specifying partial or total contents of the word rather than its location. Suppose that we have a student registration file stored in memory. Given a student #, we wish to access memory to find out whether this student is registered in any course, and perhaps get a list of courses in which he or she is registered, if any. In conventional memory organization, this will require accessing student records one by one (starting from the first location or some calculated location) until finding the desired record, or terminating the search operation at some point when the desired record is not found to exist. We may also wish to access memory to find all students registered in a particular course, or living in a given city, etc. There is no direct way to get these answers in a single access to a conventional memory module. Associative memory organization, on the other hand, will permit simultaneous access to the entire memory and enable the retrieval of the contents of the word associated with the input key or pattern without any information on the location of this word.

A schematic of a 4-word, 4-bit associative memory (containing a total of 16 storage cells) is shown in Fig. 6.7. Access to associative memory is made by supplying the input bit pattern for the key(s) to be matched. This pattern is moved to the comparand register *C* of the memory module. The match/mismatch register, *M*, has one bit corresponding to one word of memory. Register, *M*, is set to all 1s initially. Internally, the memory module uses as many cycles (strobe pulses) as the number of bits in the input pattern. During each cycle one bit of comparand register is compared with corresponding bits of all memory words. This is a form of exclusive-OR operation, and can be expressed for the *i*-th bit and *k*-th word as follows:

$$m_i \leftarrow (w_{ki}c_i' + w'_{ki}c_i)'$$

Thus every mismatch resets (to zero) register *M* in the bit position corresponding to that word. After the cycle of comparisons is completed, the contents of register M indicate the position of the word which matches the pattern of input key(s). Associative memory may also include circuitry to resolve non-unique matches.

We can select the bits of input pattern which are to be compared by introducing a mask register between the comparand register and comparison circuitry. Sometimes we may simply be interested in retrieving those memory words which have some components within a given range. For example, we may want to retrieve information on those students who are between the ages of 18–20 years. This can be achieved by setting lower and upper bound comparand register values. The nature of associative search operations can be specified by a (coded) bit pattern in an additional register known as the control input. The nature of the desired output may also take many forms. Sometimes, we may be interested in retrieving the contents of memory words (records) matching an input pattern. At other times, we may simply be interested in the total number of memory words (records) which match this pattern. To exercise this intricate control on associative memory operation, the memory module will include, in addition to data input/output and control input/output registers, a buffer area for the output to be processed by response resolution circuits as well as circuitry for over-all control of memory module operations.

In view of the above discussion it is obvious that cost per bit of associative storage is much higher than conventional random access memory. However, associative memory has provided performance improve-

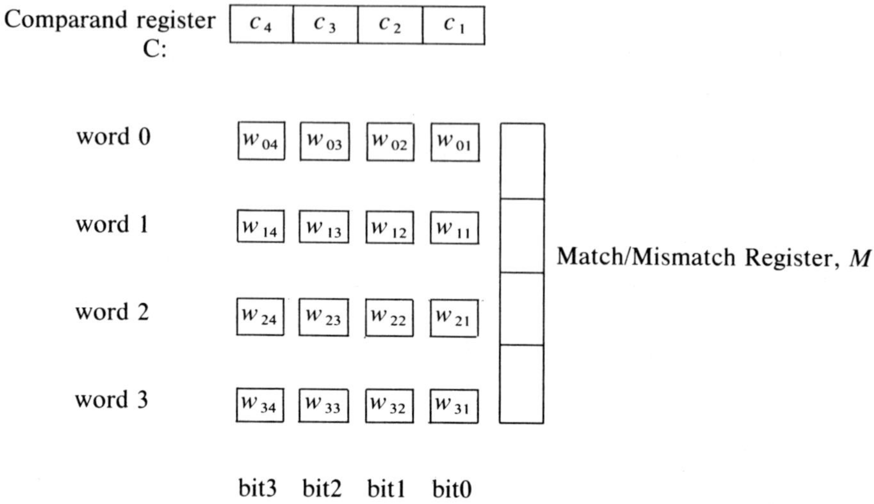

a) General Layout of associative memory with basic registers.

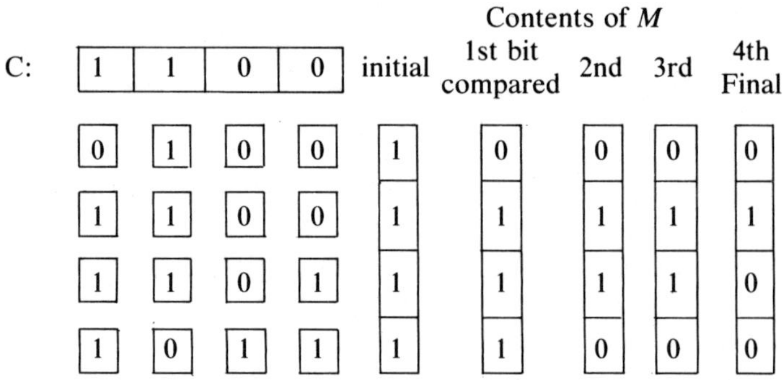

b) Examples of comparand register and memory contents and resulting contents of match/mismatch register

Figure 6.7 Associate memory

ments of many orders of magnitude for many general purpose computer systems and special systems, wherever there is a need to search tables or blocks of data based on one or more components of input patterns.

6.6 BLOCK ORIENTED RANDOM ACCESS MEMORY MODULES

In a computer system, it is desirable to keep in primary random-access memory all programs and data which are to be accessed frequently. Examples of these programs (and associated data) are monitor routines, device management and I/O transfer routines, language translators, and database management routines. Since random-access memory used for primary storage (cycle time of about 1 μsec) is expensive, the actual available memory of this type is frequently inadequate to completely satisfy the residency requirements of frequently used programs and data. Most of these programs and data must be kept on secondary storage such as magnetic drum or magnetic disk. Initiating transfers of information to or from magnetic drum or disk is very slow compared to the speed of the precessor in a computer system. This problem has been solved by attaching to primary memory some less expensive, larger capacity but slower memory module(s) known as large core storage (LCS), or extended core storage (ECS). The LCS originally produced by IBM comes in modules of one to two million bytes, has a cycle time of ~8 μsec, and cost of about ⅙ of the higher speed primary memory. This is random-access memory which means that resident programs can be executed directly. The programs which are not executed frequently (repeatedly) can be loaded in this memory and accessed by the processor at the slower speed. The programs to be executed frequently, which include most of the systems programs, can be transferred to high-speed primary memory and executed from there. The transfer rate provided by LCS is ~1 million bytes/sec. Thus, while LCS is random access memory, it is frequently used as block-oriented random-access memory. In a typical program execution, the processor may access some information from higher speed primary memory and some from LCS. The relative ratio of these accesses becomes important in measuring system performance. For example, if for every ten accesses made by the processor, nine are out of primary memory and one is out of LCS, then the average cycle time ~1.8 μsec. Another common use of LCS is to provide input/output buffers. In this case input from devices can be read into LCS for use by the processor, and output produced by the processor can be written into LCS for subsequent printing/displaying on slower devices.

The ECS produced by CDC is strictly a block oriented random-access memory. It comes in modules of 128 thousand 60-bit words, has cycle time of 3.2 μsec., and costs about ⅒ of the higher speed primary memory. The block transfer rate is 19 million 6-bit bytes/sec. Programs and data can be stored in ECS and swapped in blocks of 500-4000 bytes into primary memory whenever needed for execution. As before, ECS can also serve input/output buffering requirements very efficiently.

On the other end of the spectrum, the concept of block-oriented random-access memory can also be used by providing a small-size but very fast memory module to serve as a buffer between the processor and main random-access memory. Most processors are about 10 times faster than the primary memory speed of 1 μsec. cycle time. A buffer memory, also known as *cache memory*, can be designed to operate at a cycle time of 100 nsec. or less when some given data is needed by the processor for the first time, it will be accessed from primary memory but the block containing this data is also transferred to the buffer. On subsequent reference by the processor to this data, it is made available directly from the buffer and no access to primary memory is necessary. If we assume that four out of five accesses made by the processor will find information in the buffer (cache) then average memory cycle time as seen by the processor is 280 nsec. compared to 1μsec (1000 nsec.) of primary memory.

The buffer or cache memory can be based on associative memory design to serve other system requirements. The cache memory has been successfully used by many IBM systems and it has provided substantial improvement in system performance. Many other main-frame manufacturers have also successfully used the concept of cache or associative memory.

6.7 MAGNETIC SURFACES FOR MASS STORAGE

Instead of fabricating cores of magnetic material and threading them with wires to store bits of information, we can use a sheet of metal or plastic, coat its surface with a magnetic material, and use this surface to record bits of information by creating tiny magnets in selected regions of this surface. The polarity of magnets—the direction in which the magnet faces—determines if the bit is 1 or 0. These tiny magnets are created by placing a magnetic field near the surface as shown in Fig. 6.8. When a current is passed through the recording head, a magnetic field is generated around the gap in the head which instantaneously magnetizes the surface region immediately under the head. The direction of current in the coil will determine the polarity of the magnet on the surface. By moving the magnetic surface and by controlling the direction of current a succession of magnets (bits of 1 or 0) can be created in immediate vicinity on the surface. The magnetic material used for the surface must have high retention (of flux) properties just like the one used for magnetic cores.

For reading of information, the magnets (with their fields of magnetization) on the surface passing under the head induce a voltage whose polarity corresponds to the direction of change in the magnetic field. This induced voltage can be amplified and used for setting the contents of a buffer register. Thus the same head can be used for reading as well as

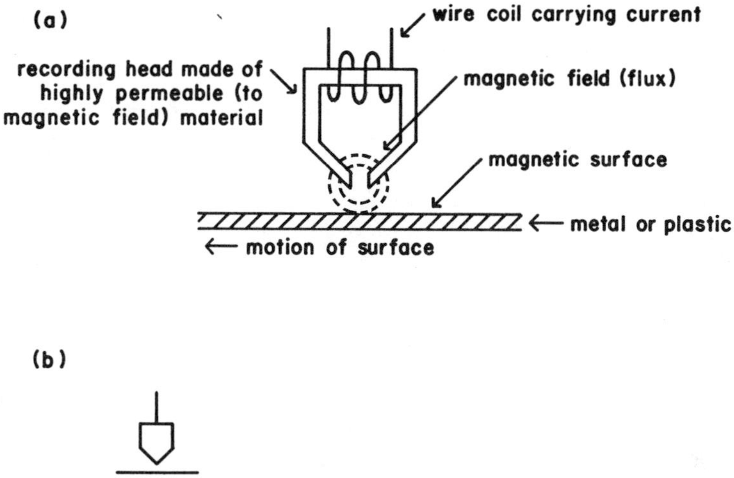

Figure 6.8 Magnetic surface for information storage
(a) Physical configuration
(b) Equivalent representation

writing of information. The number of bits recorded per inch per track may vary depending on the quality of the surface and other hardware used, up to 1600 bpi (bits per inch per track). The number of tracks on the surface depends on the device type as well as on the number of recording (read/write) heads used. A device with 8 tracks and 1600 bpi can store 1600 8-bit bytes per inch.

Some common examples of mass storage devices using the above principle are magnetic drums, discs, and tapes. Within these general categories, a variety of products can be found to satisfy user requirements. For example disc devices may consist of hard disc (like standard phonograph records) or less expensive flexible diskettes (floppy discs). Magnetic tapes are available as reels, cassettes, or continuous loop. For discs, a further variation exists in terms of whether or not the read/write head is movable across disc tracks or fixed, etc. The term mass storage used earlier in this section also needs a little explanation. The capacity of storage directly accessible to the computer without operator intervention (mounting, unmounting of discs or tapes) may range from a few hundred thousand 8-bit bytes (characters) to a few hundred million bytes or more. The actual capacity of this on-line (to computer system) storage, of course, depends on the size of computer primary memory because all information must reside here for processing. A very disportionate size of

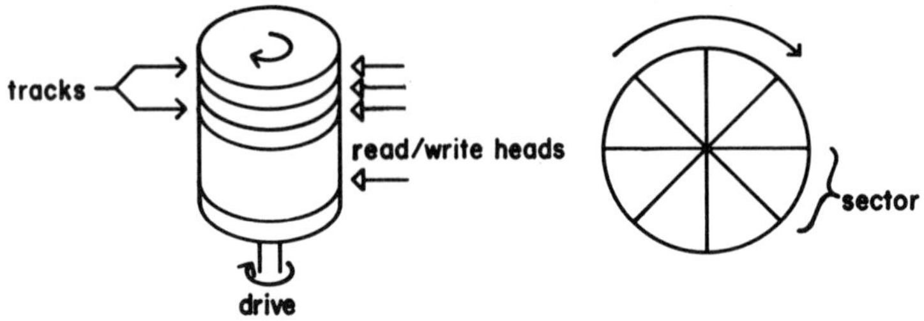

Figure 6.9 Physical layout of magnetic drum

on-line storage may create unusual traffic problems during accessing of information. The overall allowable computer system costs also limit the capability.

6.7.1 Magnetic Drum Direct Access Storage Device

Magnetic drum for mass storage medium consists of a cylinder made of high quality alloy (Nickel-Molybdenum) whose surface is coated with magnetic material. The surface consists of a few dozen to a few hundred recording tracks for storing information. During its operation, the drum is mounted on a drive and constantly rotates at a high speed (50 to 100 revolutions/sec.). There is at least one read/write head per track, and each track is divided into a number of addressable sectors. This is shown in Fig. 6.9. For a large number of closely spaced tracks on the surface of the drum, the read/write heads can be accommodated by distributing them spatially around the circumference. Since there is a fixed read/write head for each track, no mechanical movement of the heads is required. However, we do not have to perform read/write operations on all the tracks simultaneously. We can select a band of tracks electronically. This band may consist of one to eight data tracks. This is shown in Fig. 6.10.

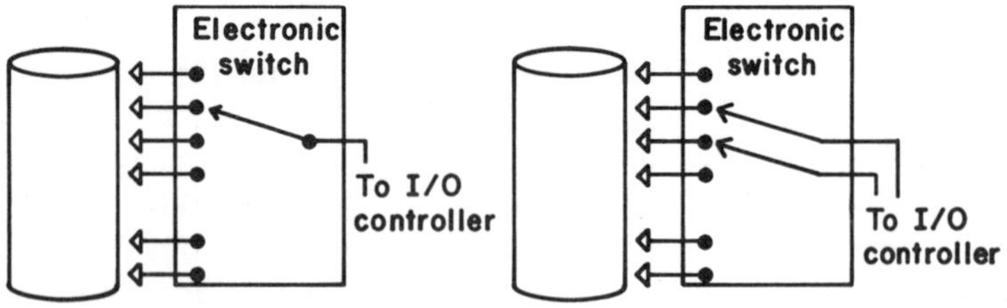

Figure 6.10 Electronic selection of band tracks for read/write operations
(a) one track per band
(b) two tracks per band

For the purpose of illustration, let us assume some typical characteristics of a magnetic drum as follows:

Total number of tracks, $m \equiv 64$
Number of tracks/band, $t \equiv 1$
Number of sectors/track, $n \equiv 128$
Time for one revolution, $r \equiv 20$ msec (1/50 sec)
Recording density, $d \equiv 1000$ bits/inch (bpi)
Circumference of drum, $c \equiv 20$ inches
Therefore, for this drum,
capacity per track $\equiv d \times c \equiv 20{,}000$ bits
Total capacity $\equiv m \times 20{,}000 \equiv 1.28$ million bits

$$\text{Data transfer rate} \equiv \frac{\text{capacity per track}}{\text{Time for one revolution}} \equiv 1 \text{ million bits/sec}$$

Suppose we had eight tracks per band so that we could record all eight bits of a byte simultaneously, then we could achieve a data transfer rate of one million bytes/sec. This is not uncommon.

Let us now consider the design of a drum storage module in a little more detail. If there are 64 bands of tracks and 128 sectors per track then we need 6 bits to address a band and 7 bits to address a given sector within the selected band for a total of 13 bits to directly access any sector on the drum. For recording information at precise intervals normally a track is reserved. On this track, a precisely spaced pattern of bits (1's) is recorded by the manufacturer. This is known as the timing track. The information on this track is used to synchronize the read/write operations so that slight variations of speed in drum rotations can be tolerated. Since the drum is constantly rotating, it must have the ability to know which sector is under the read/write heads at any given time. In fact, it must know which sector is going to be under the read/write head next so that it can initiate read/write operation immediately when the selected sector passes under the read/write head. This is achieved by reserving one track on which patterns of bits corresponding to the next sector number are recorded. This is known as the sector # track.

For read/write operations, the address of the selected sector is loaded into the address register (AR). If the operation is that of writing, then the data to be written is loaded into the buffer register (BR); the contents of this register can be shifted to write one bit at a time. The track number portion of the AR is used to select the correct head(s), and the sector number portion of the AR is constantly compared with the sector read from the sector # tracks, as brought into a modulo-128 counter register.

When the contents of the counter register and sector number in AR match, a gating signal is produced to transfer the bit(s) from BR into the head switch thus accomplishing the write operation. A read operation is similar in terms of head selection and sector match, except now data is transferred from the head to register BR. A block diagram of the circuit for the above drum operations is shown in Fig. 6.11.

While we have shown a single head to perform read and write, this head may be split where the write part leads the read part for verification of written information by internal drum controller circuitry, if desired. We may also provide more than one head around the same track for faster access.

Magnetic drum as a storage medium provides fairly high capacity (compared to primary memory), fast transfer rates, and direct access to a sector (or block of information) with an average rotational delay equal to the time for a half revolution.

6.7.2 Magnetic Disc Direct Access Storage Device

Magnetic discs are made of plastic and both sides of the surface are generally coated with magnetic material for recording of information (like phonograph records). One or more (generally about 10-12) discs are fixed on a spindle. This disc assembly is mounted on a drive so that the entire assembly can be rotated at a high speed. Each disc surface consists of a large number of recording tracks. Read/write heads can be permanently

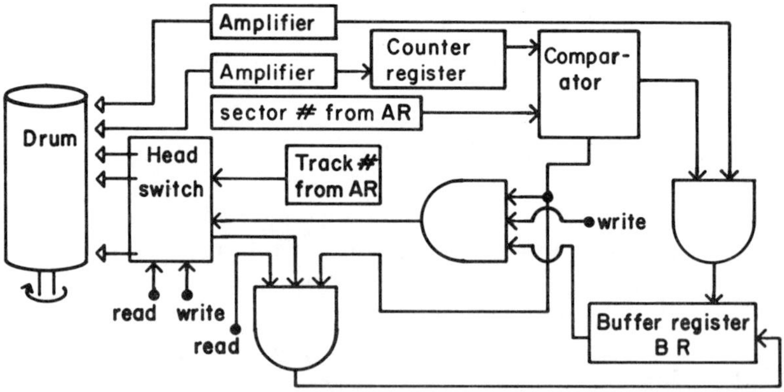

Figure 6.11 Drum storage module circuit configuration. Amplifiers are used to increase the magnitude and shape the pulse generated by flux changes sensed by read heads. Similarly drivers (not shown) are needed to produce drive currents for inducing flux changes in the write heads

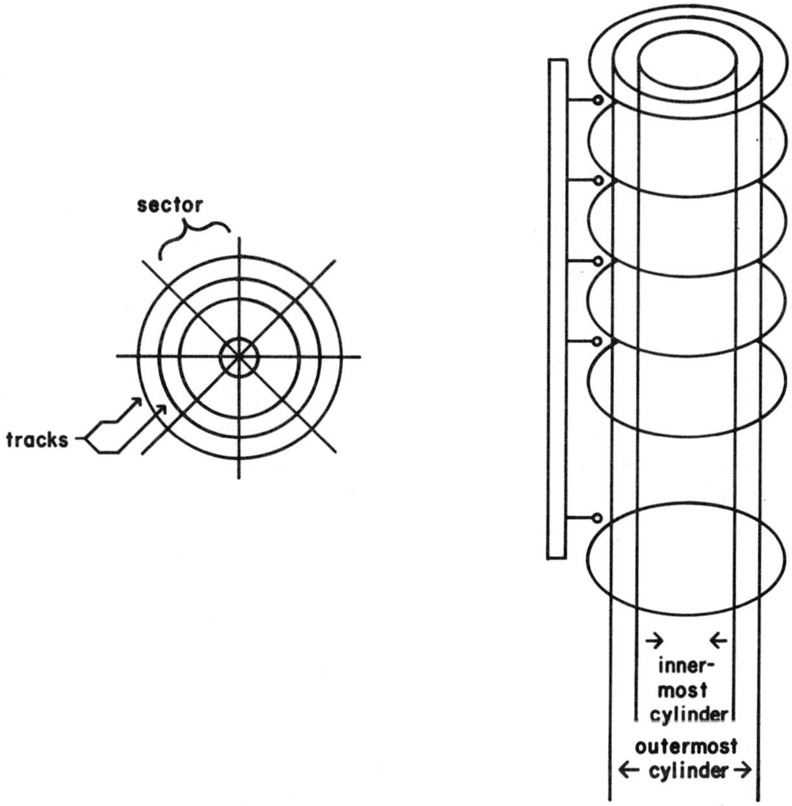

Figure 6.12 Disc device configuration

positioned at each track in which case the disc assembly behaves like a set of concentric drums called a fixed-head disc, or the read/write heads are made to move to the desired tracks called a moving-head disc. The basic configuration of disc devices is shown in Fig. 6.12. The entire assembly of read/write heads is known as the access mechanism. There are usually as many read/write heads as the total number of surfaces (excluding the upper surface of the top disc and low surface of the bottom disc). For a fixed head disc there is one set of read/write heads for each track of the cylinder and the access mechanism does not move to seek the desired track. Since the total surface available for storage (number of tracks per cylinder X number of cylinders) is much higher, disc devices of up to a few hundred megabyte capacity have been produced. The only drawback in the operation of a moving head disc is the seek time to position the heads on the desired cylinder tracks. This time is typically 50 msec. or higher. The rotational delay in accessing a selected sector on a track is similar to drum, although the magnitude of this delay is somewhat higher, about 25 msec. or more. The slower rotational speed of discs is due to their mechanical construction.

For low-cost small computer systems, soft discs known as floppy discs or diskette devices are available. They provide an inexpensive means of attaching up to one megabyte of on-line direct access storage per diskette. By attaching a number of drives to the computer, the on-line storage capacity can be increased substantially. Depending upon the disc drive, the disc itself may be fixed in an enclosure known as a disc pack to protect the disc surface from damage.

The disc device provides fairly high data transfer rates, typically 100-500 kilobytes per sec. The seek time and rotational delay can both be minimized through a combination of software, hardware, or firmware support for controlling input/output. For example, the positioning of the access mechanism on a track can be continually maintained in a manner such that very little or no movement is required for the next I/O request.

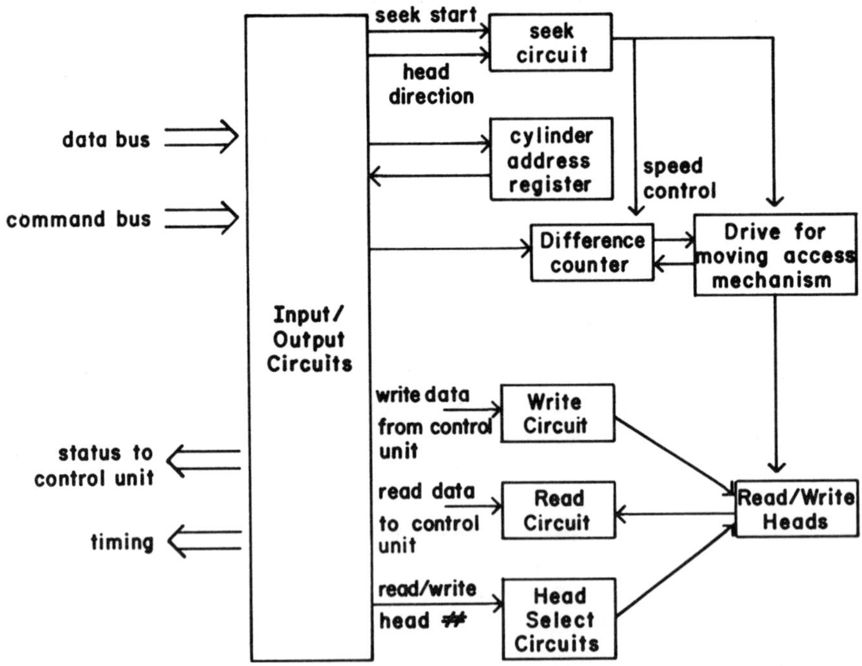

Figure 6.13 Typical configuration of functional circuits for disc operation

We can keep track of available or free storage areas on the entire device. When a write operation is issued, the information can be written on the next available sector to pass under the read/write head rather than on a specified sector. These techniques require keeping a directory of all information stored on the disc, and controlling input-output requests through this directory rather than under the discretion of user programs.

The operation of disc storage devices will now be explained with the help of a functional block diagram shown in Fig. 6.13.

Let us assume that the disc is rotating at proper speed and the access mechanism is currently positioned at some track (cylinder), x. Let the address of the desired track (cylinder) be y. The magnitude of the difference, $|x - y|$, is sent to the difference counter, the new cylinder address y is placed in the cylinder address register, and the sign of the difference is used to determine the direction of movement of the access mechanism which is supplied to the seek circuits. Depending on the number of tracks (cylinders) to be traversed, the seek circuits will also be signaled to move the access mechanism at high speed initially or traverse the whole distance at moderate speeds. When the desired cylinder is reached, an attention signal is sent to the control unit. Head-select circuits are then enabled by the control unit followed by the actual transfer of input/output data.

6.8 SOLID STATE "DISC" MEMORY MODULES

The mechanical aspects of moving magnetic memories such as drum or disc not only result in substantial latency time (seek time plus rotational delay) but also require regular maintenance. The drum and the disc have read/write heads close to the surface which is moving at high speeds. Both the surface and the heads can gather dust requiring cleaning. The surface may also be scratched causing permanent damage. A limited number of spare tracks may be provided on expensive devices. Any technique which eliminates the movement of the surface and still provides large storage capacity is considered highly desirable. For mass storage applications, the principle of accessing a large block of stored information (sector address), rather than a word or few bytes, is still valid. Two strong contenders to moving magnetic storage which are becoming popular because of their superior performance are charge coupled devices (CCD) and magnetic bubble memory (MBM). CCD uses MOS technology but is based on capacitive charges rather than an aggregation of transistors as storage elements. MBM uses magnetic technology but is based on bubbles (magnetic dipoles) created in the material rather than surface magnetization. With properly applied voltage to a CCD contact a charge can be

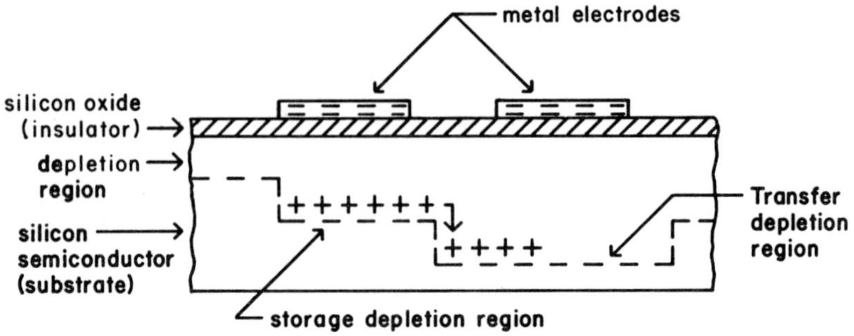

Figure 6.14 Operation of CCD

stored as a 1-bit and by applying proper voltages to a succession of contacts this charge can be made to move. Similarly, a magnetic field when applied to a region of suitable magnetic material can create a bubble (1-bit), and the bubble can be moved in the desired direction under magnetic field gradient. Both of these technologies can provide a high density of storage elements, a few megabits/in^2. A single chip can be produced with 50 to 100 kilobits or higher, with a group of a few hundred to a few thousand bits read or written in one access. We will now discuss the details of operation of these devices.

MOS processing is used to fabricate low-power dynamic-shift registers known as CCD. The CCD consists of MOS capacitors, capable of producing and storing minority carriers below the interface of semiconductor and oxide layer, as shown in Fig. 6.14.

The stored charge can be moved along the surface of the semiconductor in either direction by application of the proper voltage levels to metal electrodes in the desired sequence. For example, if a bias voltage is applied to the metal electrode it creates a depletion zone at the interface of the semiconductor substrate and the metal oxide insulator (shown as the threshold region). Next, when a sufficiently larger voltage, known as the "storage" potential, is applied to the same electrode, it creates a well which accumulates mobile charge (minority carriers). This represents a stored binary value of 1 at the MOS capacitor corresponding to the electrode to which this storage potential is applied. If a larger voltage, known as the "transfer" potential, is applied to the next electrode, it deepens the well resulting in the transfer of charge. Once the charge has been transferred, the voltage on the first capacitor is returned to the normal biasing potential, and the voltage on the receiving capacitor is lowered to the storage potential. Thus an aggregate of MOS capacitors produced by integrated circuit technology form a shift register capable of storing bit patterns, shifted in for storage, and shifted out for reading.

Different techniques have been employed for electrode connections, clocking of potential and refreshing of charges due to leakage in capacitors. One of the commercially available CCD memory integrated circuit chip is illustrated in Fig. 6.15.

Magnetic bubble memory is constructed from thin films of synthetic ferrites or garnets with desirable characteristics such as small bubble size to allow high density of storage, high mobility of bubbles under applied magnetic field facilitating access of stored information, and good temperature characteristics providing reliability of operation. In the absence

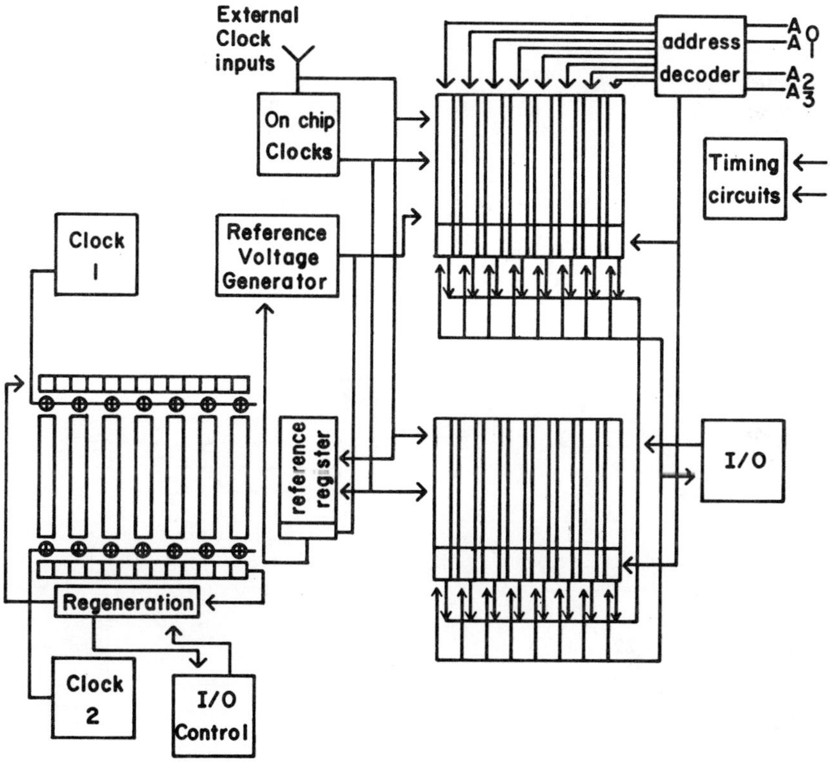

Figure 6.15 Organization of Texas Instruments TMS3064 64-kilobit CCD memory
(a) One 4 kilobit register (loop)
(b) Overall chip organization containing 16 4K loops

of a magnetic field, a thin wafer made from these materials will posssess serpentine domains of opposite magnetization occupying equal areas as shown in Fig. 6.16(a). These domains can be reduced to small bubbles representing stored 1 bits under a strong magnetic field.

In practical MBM devices, suitable techniques have been devised to control generation, propagation, transfer, detection, replication, storage and annihilation of magnetic bubbles. An example of the organization of a commercially available 92-kilobit chip using MBM is shown in Fig. 6.17 In this case, there are 157 registers and each register contains 641 bubble positions giving a total storage capacity of 157 × 641 bits. The bubbles in each register are moved around a loop for accessing information. These registers are called minor loops. The bubbles to these minor loops are transferred from a major loop for storage of bit patterns, and are transferred to the major loop for reading.

In practice, not all of the minor loop registers are used for storage. For example, only 144 out of 157 are available at any given time, the other 13 are spares to offset defective ones.

To access information stored in minor loops, the bubbles are moved around the loop until the desired bit position (bubble) in all loops is at the top position. At this time the transfer element is activated causing removal of bubbles from the minor loop and entering them into the major loop. The bubbles are then shifted to the detector section. If destructive read is desired then the replicator is set to deflect the bubble into the detector

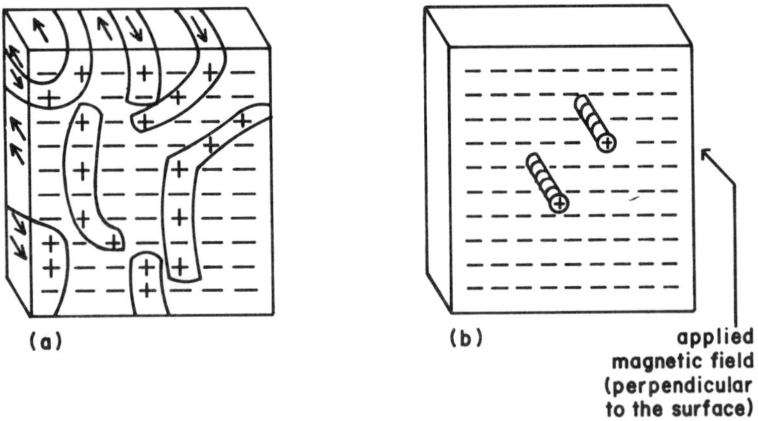

Figure 6.16 Domains in magnetic material
(a) Serpentine domains in the absence of magnetic field
(b) Shrinking of serpentine domains to cylindrical domains (bubbles) under large magnetic field

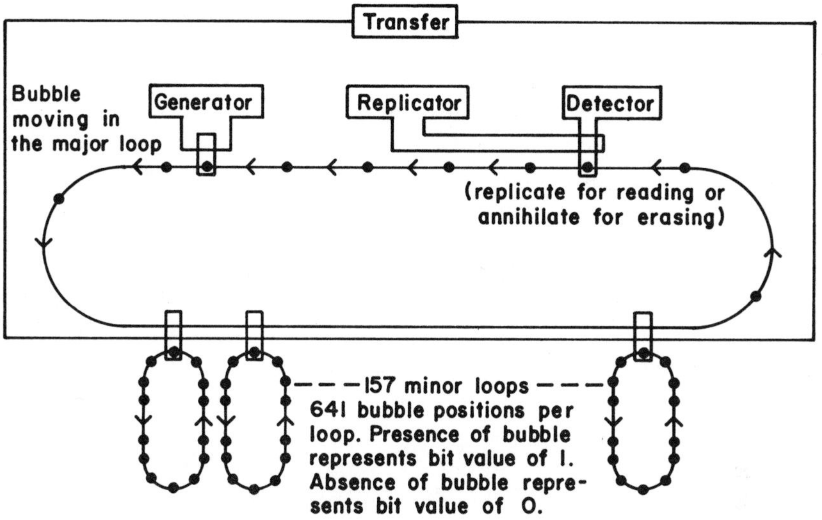

Figure 6.17 Magnetic bubble memory chip organization

track, otherwise the replicator will produce a copy of the bubble and enter it into the major loop. To store information, data is entered into the major loop via the generator current loop. The creation of a bubble represents a bit value of 1, its absence represents a bit value of 0. A block of 157 bits are entered this way for subsequent storage into the specified position of all minor loops. When all bits (bubble positions) have been generated and shifted to (positioned at) the corresponding minor loop positions, the transfer element is activated as before. Note that during the read operation, bubbles were transferred out of the loop positions of minor loops. At the end of major loop cycle, they will be transferred back for retention of stored information.

Both the CCD and MBM devices are becoming popular for secondary storage where block oriented transfer of information is desired. Many of the word processing terminals, particularly the portable ones are being provided with MBM memory modules for entering and storing data which may be transferred later to a computer for detailed processing. Research work at the Bell Laboratories served to demonstrate the basic concepts related to CCD and MBM as early as 1970. Practical development of these devices took several years of laboratory design work. First commercially available single chip 64-Kbit CCD and 92-Kbit MBM were announced in 1977. Further improvements in the design and fabrication techniques are expected to push these devices in the multimegabit range.

Some of the factors influencing the acceptance of CCD and MBM devices within the computer memory hierarchy have been the reliability of operation, the cost per bit, and the access time. Earlier versions of CCD and MBM chips suffered from technical deficiencies such as loss or alteration of information. More recent versions have overcome these problems although even now they may not fully satisfy military specifications for field operation. In terms of end-user system cost, CCD and MBM devices provide substantial cost reduction over MOS and fixed-head disk or drum technologies within 500-Kbit to 5-Mbit range. Floppy disk storage costs are comparable. Access time of CCD and MBM devices is higher than Cores/MOS devices but substantially lower than disk storage. Price and access time comparisons to the end user for a *1-Mbit* storage capacity are summarized below:

	Price/bit (cents)	Access time (μsec)
Cores/MOS	1	.1–1
CCD/MBM	.1	100–1000
Drum/disk	1	10,000–50,000
Floppy disk	.1	100,000–500,000

From this summary, it is clear that CCD and MBM devices are likely to receive greatest acceptance at the middle level of computer-memory hierarchy between the expensive but faster Cores/MOS primary memory and cheaper but slower moving magnetic memory.

PROBLEMS

6.1 Determine the total number of drive and sense amplifiers required for a 1,024 word, 16-bit memory based on (a) linear selection, and (b) coincident selection.

6.2 Give examples of volatile and non-volatile storage elements used in the design of memory modules. How is the stored information properly retained in volatile storage elements?

6.3 Read-only-memory (ROM) modules are being increasingly used in the organization of digital computers. Give a list of ROM applications. ROM used in digital systems may be non-alterable or electrically alterable with the ability to erase stored information and rewrite new information. How is the stored information erased? How is the new information stored? Under what conditions would the use of erasable ROM be more desirable?

6.4 Discuss the advantages/disadvantages of dynamic storage versus static storage module.

6.5 For a two-step memory hierarchy, the average access time, neglecting second order terms, is

$$t = f(1 + \alpha r)$$

where f is the access time for fast memory,
r is the ratio of access time of the slow memory to that of the fast one, and
α is the fraction of all accesses that must be made to the slower memory.

State the conditions under which the two-step hierarchy scheme will approach the speed of its fast member. Give examples of how these conditions may be realized in practice.

6.6 A magnetic drum has a 20-in. diameter, a bit density of 300 bpi, 64 tracks, and a rotational speed of 4000 rev./min.
(a) What is the storage capacity in bits for this drum?
(b) What is the data transfer rate?
(c) What is the expected access time if there is one read/write head per track?
(d) What is the data transfer rate if 4-tracks are read or written simultaneously?
(e) If there are 32 sectors per track, how many bits are required to address any sector?

6.7 A magnetic disc assembly consists of 12 discs, 202 tracks per disc; rotational speed of 2400 rev./min., and 8000 bytes per track.
(a) What is the storage capacity per cylinder? What is the storage capacity of the module?
(b) What is the data transfer rate?
(c) Assume an expected seek time of 50 msec., what total time must elapse before transfer of information can take place?

6.8 When disc storage is compared to drum storage, the major slowing of transfer operations in disc occurs due to seeking of the desired cylinder. How can this seek time delay be reduced or entirely eliminated? Suggest practical methods of achieving this goal.

6.9 Magnetic reel or cassette tapes are used extensively as an inexpensive medium for mass storage applications. Give examples of the use of these media in computer systems. Describe the operation of tape devices in terms of access time and transfer rate.

6.10 Solid state secondary storage devices such as CCD and MBM offer an attractive alternative to moving magnetic memory devices, but have not replaced them yet. Why? Are there other mass storage devices which are likely to compete with these in the near future? Give examples and discuss their potential.

REFERENCES

1. Brooks, F. P. Mass Memory in Computer Systems. *IEEE Transactions on Magnetics*, Vol. MAG-5, No. 3, 1969, pp. 645–639.
2. Flores, I. *Peripheral Devices*. Englewood Cliffs, N.J.: 1973.
3. Hnatek, E. *A User's Handbook of Integrated Circuits*. New York: John Wiley & Sons, 1973.
4. Sharpe, W. F. *The Economics of Computers*. New York: Columbia University Press, 1970.
5. Smith, A. B., ed. *Bubble Domain Memory Devices*. Dedham, MA: 1974.
6. Toombs, D. An Update: CCD and Bubble Memories. *IEEE Spectrum,* vol. 15, no. 4, 1978, 22–30.

Chapter 7

ARITHMETIC/LOGIC UNIT AND CONTROL UNIT

One of the functional modules making up a computer system is the central processing unit (CPU) that controls both the transmission and transformation of information within the computer in a way determined by the sequence of instructions stored in memory unit. The operations of the CPU are normally done in sequential time units known as instruction cycles. In a typical instruction cycle, the following operations are done:

1. An instruction is fetched from the memory unit.
2. The fetched instruction is decoded.
3. Operand(s) are fetched from the memory unit if so required.
4. The operation specified in the instruction is performed.
5. Control is returned to (1.).

As described in Chapter 1, the CPU is generally subdivided into two functionally different units, the arithmetic/logic unit (ALU) and the control unit (CU) which may or may not be implemented as separate modules. The ALU is used for information transformation and the control unit is responsible for the generation of all the control signals. The control unit is generally the most complex part of a computer and may be looked at as the brain of the computer. There are in general two different approaches which can be taken to realize a control unit, a hardwired approach and a microprogrammed approach although in many cases, it is difficult to draw a clear separating line between them and to obtain designs employing strictly only one of the two approaches. In Chapter 4, various arithmetic, logic, and sequencing circuits have been described. In Chapter 5, the organization and interconnection of various functional modules in a computer were briefly described. We are now in a position to discuss in more detail the design of the CPU.

Instruction	Length (bytes)	Description
LDI	2	load AC with immediate data
SLL	1	shift content in AC left logically
LDA	3	load AC
ADD	3	add to AC
STA	3	store AC content
ISZ	3	increment and skip three bytes if zero
JMP	3	unconditional jump

Table 7.1 The subset of chosen instruction.

Rather than looking at the entire instruction set of a computer, we choose a representative subset of an instruction set found on a typical minicomputer or microcomputer. Note that even in a typical microprocessor, the number of instructions is seldom less than 40 and that a typical larger computer has over a hundred instructions employing various sophisticated addressing schemes. However, our major concern here is to show the design techniques and not to produce a typical working design. The subset of chosen instructions is shown in Table 7.1. The chosen instructions have very simple forms. No indexing or indirect addressing has been included to simplify the subsequent discussion. Also, no explicit I/O instructions are included. The reader should have no difficulty in

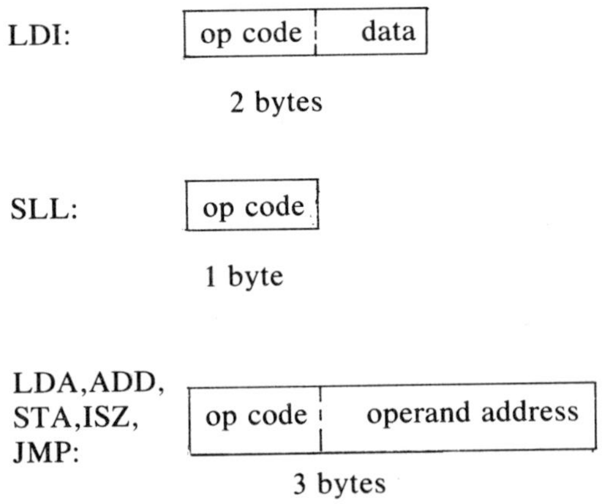

Figure 7.1 Instruction formats

subsequently extending the scope of the chosen instructions. The formats of the instructions are shown in Fig. 7.1.

It is important to realize that the instruction set chosen has a great effect on the power as well as the characteristics of the ALU and the overall computer system. Instruction formats can be 0-address (utilizing a stack feature), 1-address, 2-address, 3-address, and even 4-address (the 4th address being the next instruction address) implying different instruction word lengths. The addresses are used to reference source and destination operands (if any) used in the operations. Given the operational specifications that the eventual computer is to possess and the applications for which it is intended, the first step in the design of a CPU is usually the specification of an appropriate instruction set. After the instruction set has been specified the design of circuits for instruction execution can commence.

7.1 ARITHMETIC CIRCUITS

Before we look at the overall design of the CPU, we take a closer look at one of the major components, the ALU. The basic set of arithmetic operations that a computer is required to perform consists of addition, subtraction, multiplication, and division. A general discussion of the ALU has been given in Chapter 4. Here we will show that depending on the particular operational requirements, a wide range of implementation methods can be employed to achieve the arithmetic operations. Some of the more conventional methods are now described below.

7.1.1 Addition and Subtraction

Addition and subtraction are normally achieved using a single-addition circuit. This can be done when negative numbers are represented in 2s complement form in which case, subtraction of two numbers is simply the addition of the minuend and the 2s complemented form of the subtrahend.

When the bits of two numbers to be added are available at the same time, a set of n 1-bit full adders can be cascaded together to form an n-bit adder. This is illustrated in Fig. 7.2. This cascaded adder is sometimes called a *ripple-carry adder* for the obvious reason. Design of 1-bit full adders has been considered in Chapter 2. For addition of two n-bit numbers, a total of n full adders are required. Overflow occurs when both operands are positive and the sum bit S_{n-1} is 1 whereas underflow occurs when both operands are negative and the sum bit S_{n-1} is 0. The imple-

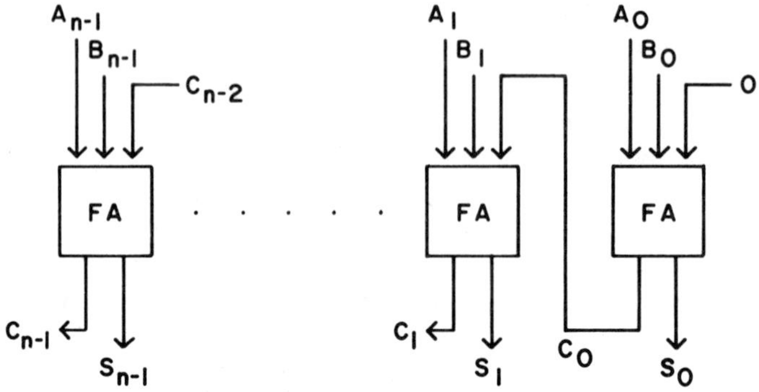

Figure 7.2 Cascaded adder

mentation of an overflow and underflow checking circuit is straightforward and will be left to the reader.

The operational speed of the above ripple-carry adder is limited by the total delay needed for propagation of the carry bits from the less significant bits to the more significant bits. Cascaded adders in which look-ahead features are incorporated have been designed to increase their operational speeds. A general k-stage carry look-ahead adder is shown in Fig. 7.3. In this circuit, the carry bits are propagated only from one look-ahead unit to the next higher order unit. Within each look-ahead unit, all the $2k$ inputs are used to produce each of the k sum bits and the single carry bit. Notice that when $k = 1$, we have the ripple-carry adder.

Under less stringent speed requirements and when bits are available sequentially, it is possible to obtain an adder circuit using only one single full-adder. A serial adder of this type is shown in Fig. 7.4. The flip-flop, FF, is used to temporarily store the carry bit to be used in the next bit

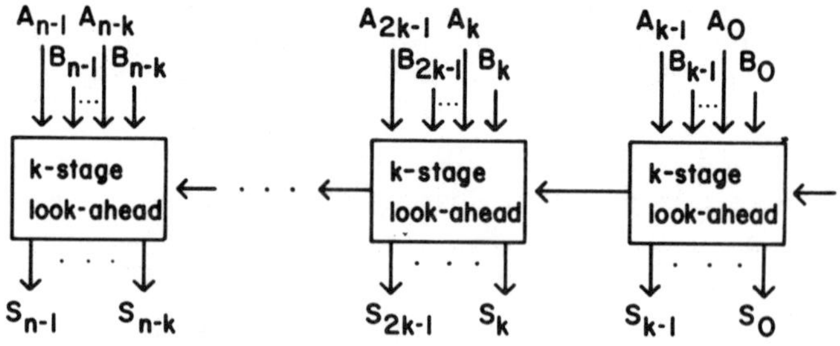

Figure 7.3 A k-stage look-ahead adder

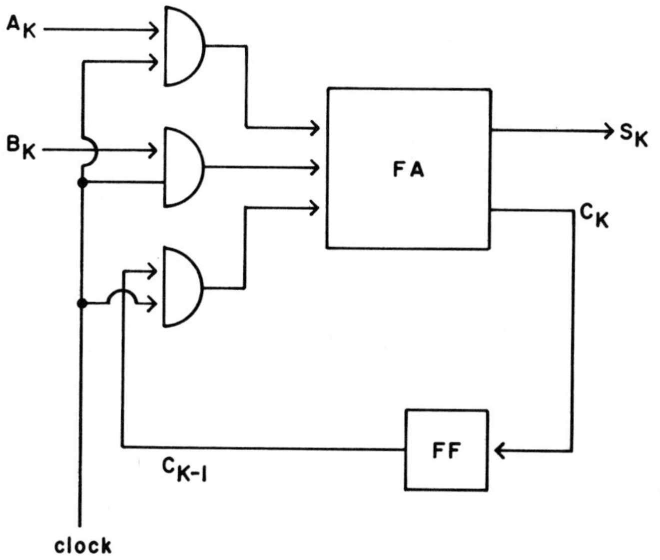

Figure 7.4 Serial adder

addition stage. In general a counter is required to keep track of the number of stages needed to perform the serial addition of two n-bit numbers.

7.1.2 Multiplication and Division

Multiplication can be accomplished with an adder through the use of the so-called shift-and-add method similar to that used in decimal arithmetic. A circuit that can be employed to perform multiplication on two n-bit numbers using this method is shown in Fig. 7.5. Both the multiplier and the multiplicand are assumed to be positive numbers. It is, of course, possible to directly deal with negative numbers expressed in 2s complement form by incorporation of additional correcting or adjusting subcircuits. The multiplier is assumed to be initially in the n-bit register B with the least significant bit (LSB) to the right end. The multiplicand is in register C which is also n bits long. Depending on the value of LSB (0 or 1), the number in register A (n bits long) and that in C are added (LSB = 1) or not (LSB = 0). The partial product in the register pair A - B is then shifted one place to the right and the process repeated. After the appropriate number of shifts, the final product will be in the register pair A - B.

The above multiplier circuit operates on each bit of the multiplier serially, thus requiring as many shift-and-add operations as the length of

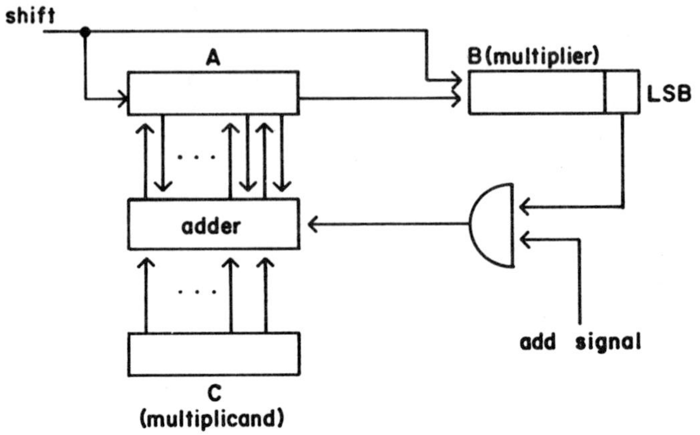

Figure 7.5 Multiplier circuit

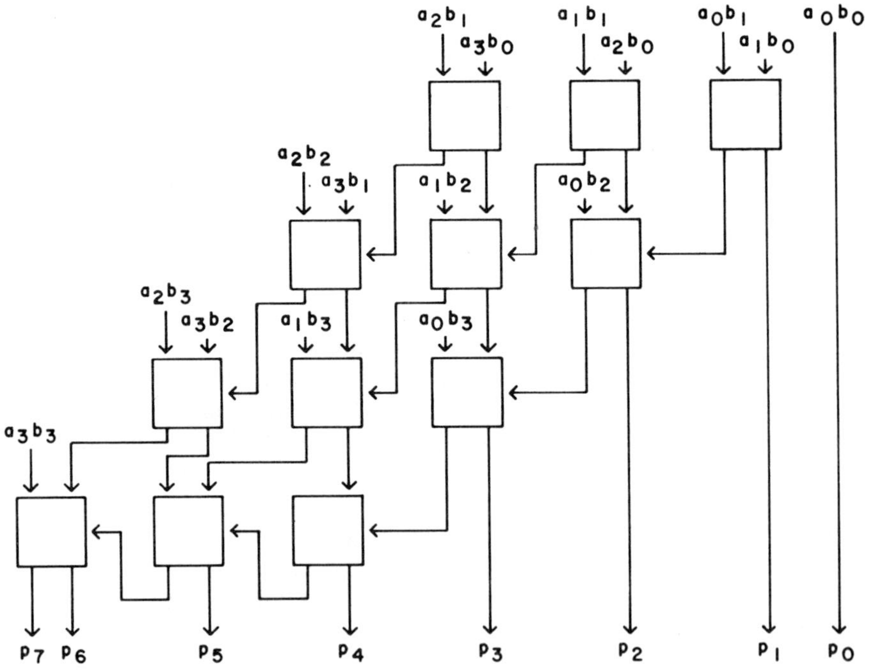

Figure 7.6 Parallel multiplier

	Accumulator register	Quotient register	
	0 0 0 0 0 0 0 0	1 0 0 0 1 1 0 0	
Shift	0 0 0 0 0 0 0 1	0 0 0 1 1 0 0 0	
Subtract	0 0 0 0 1 1 1 0		
Negative	1 1 1 1 0 0 1 1	0 0 0 1 1 0 0 0	
Shift	1 1 1 0 0 1 1 0	0 0 1 1 0 0 0 0	
Add	0 0 0 0 1 1 1 0		
Negative	1 1 1 1 0 1 0 0	0 0 1 1 0 0 0 0	
Shift	1 1 1 0 1 0 0 0	0 1 1 0 0 0 0 0	
Add	0 0 0 0 1 1 1 0		
Negative	1 1 1 1 0 1 1 0	0 1 1 0 0 0 0 0	
Shift	1 1 1 0 1 1 0 0	1 1 0 0 0 0 0 0	
Add	0 0 0 0 1 1 1 0		
Negative	1 1 1 1 1 0 1 0	1 1 0 0 0 0 0 0	
Shift	1 1 1 1 0 1 0 1	1 0 0 0 0 0 0 0	
Add	0 0 0 0 1 1 1 0		
Positive	0 0 0 0 0 0 1 1	1 0 0 0 0 0 0 1	
Shift	0 0 0 0 0 1 1 1	0 0 0 0 0 0 1 0	
Subtract	0 0 0 0 1 1 1 0		
Negative	1 1 1 1 1 0 0 1	0 0 0 0 0 0 1 0	
Shift	1 1 1 1 0 0 1 0	0 0 0 0 0 1 0 0	
Add	0 0 0 0 1 1 1 0		
Positive	0 0 0 0 0 0 0 0	0 0 0 0 0 1 0 1	
Shift	0 0 0 0 0 0 0 0	0 0 0 0 1 0 1 0	
Subtract	0 0 0 0 1 1 1 0		
Negative	1 1 1 1 0 0 1 0	0 0 0 0 1 0 1 0	END

Table 7.2 Non-restoring division.

the multiplier. Although the average number of shift-and-add operations can be reduced through a modification of this basic scheme, e.g., shifting through consecutive 0's or 1's with no intermediate additions performed, it is often desirable or even necessary to have high-speed multipliers, especially in the design of special-purpose computers in application areas like digital signal processing.

A 4-bit multiplier circuit which operates on all the four bits of the multiplier in parallel is shown in Fig. 7.6. The circuit uses an array of FAs

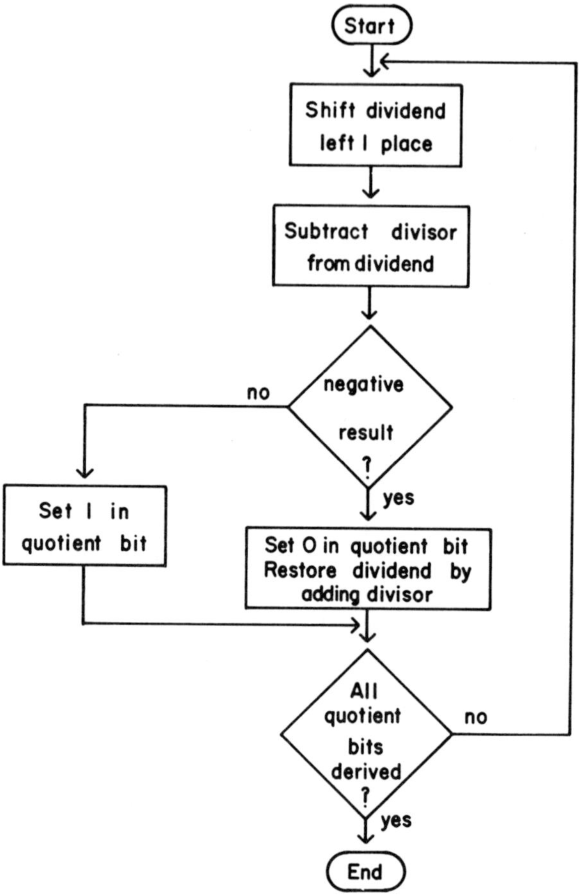

(a) Flowchart for restoring division

Figure 7.7 Division algorithms

and HAs (half-adders) to perform summation of the product terms which are generated simultaneously through the use of a set of AND gates. As the widths of the two numbers to be multiplied get large, the number of FAs required increases rapidly so that at some point, this parallel multiplier becomes economically unfeasible to build. One solution which has been adopted in a number of computers is to break up the multiplier into a number of groups of bits and to perform multiplication of the multiplicand by each group one at a time followed by an appropriate shift op-

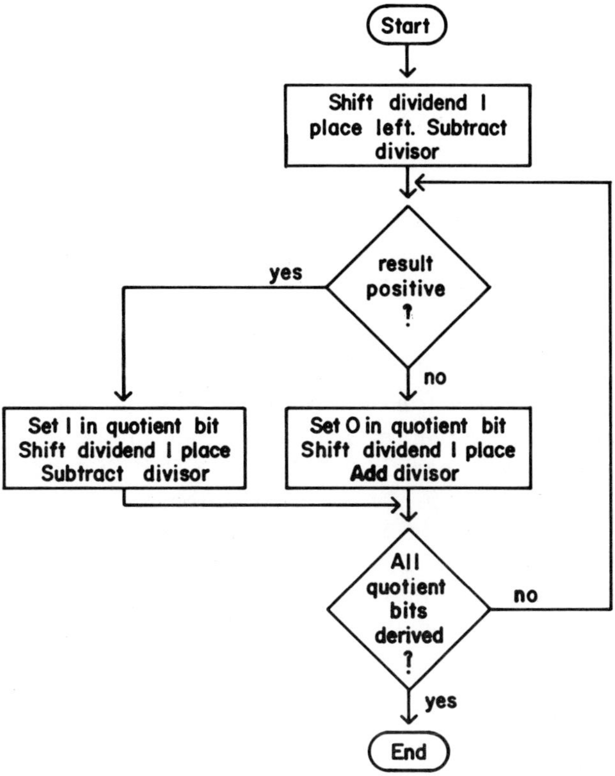

(b) Flowchart for non-restoring division

eration. This method is thus a combination of the serial and parallel methods.

Division can be similarly accomplished through a shift-and-subtract method. The shift-and-subtract is essentially a trial-and-error method and can be implemented in a number of ways. The two most popular ones are the restoring and non-restoring algorithms, flowcharts of which are shown in Fig. 7.7. In the restoring algorithm, after each shift operation, the divisor is always subtracted from the remainder. If the result is negative, indicating an error, the divisor is added back or restored and the corresponding quotient bit is then set to 0; otherwise the quotient bit is set to 1. In either case, the next shift-and-subtract loop is entered. On the other hand, in the non-restoring algorithm, in the case of an error in the present loop, the correction is taken care of in the next loop. An example on the use of the nonrestoring algorithm is given in Table 7.2where decimal 140 is divided by decimal 14 giving a quotient of decimal 10. Other methods used to perform division include table look-up and multiplication of the dividend by the reciprocal of the divisor expressed in the form of a series.

In low-cost microcomputer systems, where the ALU only has the capability of performing ordinary addition (and subtraction in 2s complement form) multiplication and division are exclusively done in software. Only in very special applications are hardware multipliers and division circuits available to achieve high speeds.

7.2 BASIC STRUCTURE OF A CONTROL UNIT

We now examine the structure of a CPU, a block diagram of which is shown in Fig. 7.8. The control unit (CU) issues all the necessary signals to control the various operations within the CPU. These signals include those that are needed to perform various arithmetic and logic operations on values in the accumulator AC and the temporary buffer register, testing of one or more of the status flags discussed in Chapter 4, and manipulation of the contents in the data counter (DC), and the program counter (PC). All these are generated from the CU through the decoding of an instruction located in the instruction register (IR). Information taken from or directed

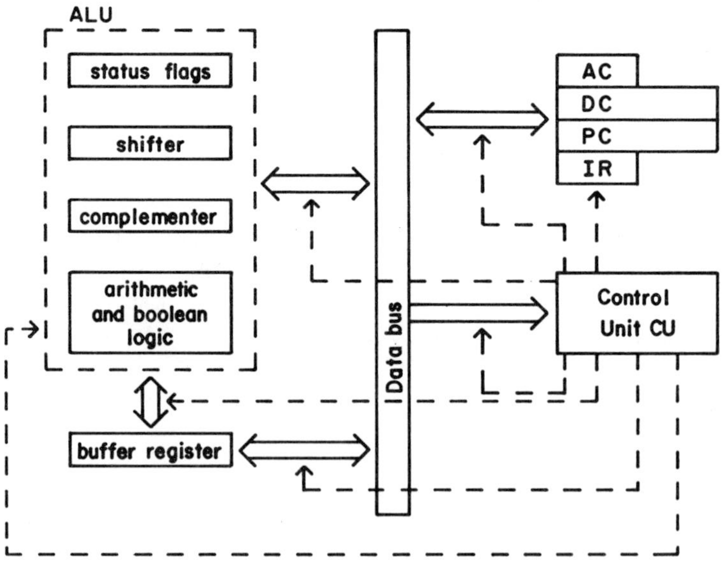

Figure 7.8 The CPU: Internal Structure

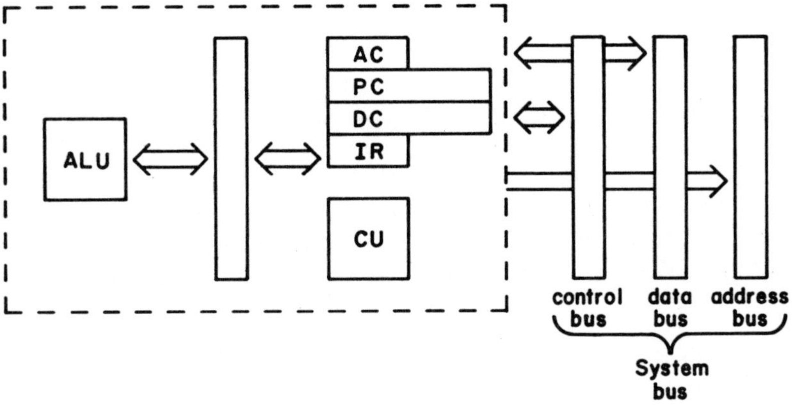

Figure 7.9 The CPU with the System Bus

to units outside the CPU is transmitted through a system bus as shown in Fig. 7.9. As has been shown in Chapter 5, the system bus is essentially made up of three sets of buses:

1. The control bus which conveys information needed to specify the types of transfer to be done, like read/write (R/W) signal, interrupt request, and so on.
2. The data bus for transferring data operands to and from the CPU.
3. The address bus for transferring addresses of either data operands or instructions to the corresponding storage devices.

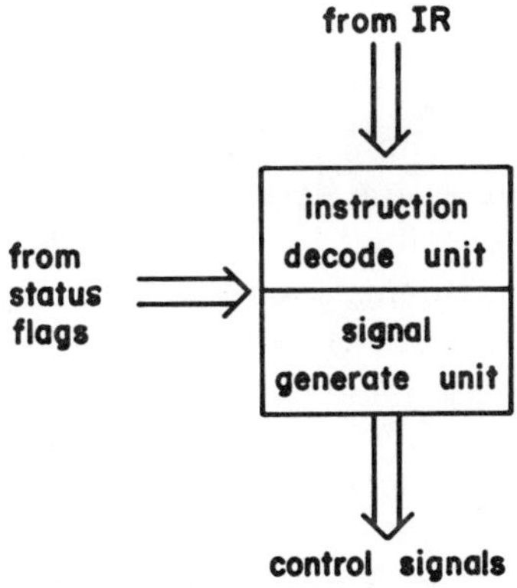

Figure 7.10 Structure of Control Unit

The structure of the CU is shown in block diagram form in Fig. 7.10. The instruction in IR is decoded in the decode unit and together with additional information coming from the status register, enables control signals to be generated by the signal generate unit. The period taken to execute an instruction constitutes one instruction cycle. However, certain instructions take a longer time to execute than others and so we divide an instruction cycle into machine cycles. In most CPUs, a machine cycle is of a fixed duration usually made up of a fixed number of states whereas in certain CPUs, a machine cycle, depending on the particular type of instruction involved, can be made up of differing numbers of states. In the design description to follow, each machine cycle will be assumed to be of a fixed duration although the reader can easily extend the description to include cases involving variable length machine cycles.

To simplify the presentation, interrupt features required for achieving the cooperation of system components will be ignored here and will be considered in a later chapter. We first look at the design of a hardwired control unit and then at the design of a microprogrammed control unit.

7.3 HARDWIRED CONTROL

Consider the subset of instructions given in Table 7.1. Each instruction cycle is essentially made up of a series of read and write machine cycles which are defined as follows:

1. A read-machine cycle is made up of a read operation followed by the necessary execution operations to be performed as will be explained later in the section.
2. A write-machine cycle is made up of a write operation (onto external devices or main memory).

All the instructions in Table 7.1 can be executed through the concatenation of one or more of the above two types of machine cycles. For example, LDI can be executed by the use of two consecutive read-machine cycles, the first one being used to access memory to obtain the first byte of the 2-byte instruction which contains the operation code and the second one being used to access memory to obtain the second byte of the instruction which contains the immediate data to be loaded into AC. The two types of machine cycles can be distinguished by means of a read/write (R/W) control line. For example, if we use $R/W = 1$ for a read cycle and $R/W = 0$ for a write cycle, Table 7.3 shows the cycle by cycle operation of the set of instructions. If indexing or base addressing is allowed, it is

Instruction	# of Cycles	Cycle	R/W	Description
LDI	2	1	1	Fetch op code
		2	1	Fetch immediate data and load into AC
SLL	2	1	1	Fetch op code
		2	1	Logically shift AC left
LDA	4	1	1	Fetch op code
		2	1	Fetch high-order operand address
		3	1	Fetch low-order operand address
		4	1	Fetch operand and load into AC
ADD	4	1	1	Fetch op code
		2	1	Fetch high-order operand address
		3	1	Fetch low-order operand address
		4	1	Fetch operand and add
STA	4	1	1	Fetch op code
		2	1	Fetch high-order operand address
		3	1	Fetch low-order operand address
		4	0	Store data
ISZ	5	1	1	Fetch op code
		2	1	Fetch high-order operand address
		3	1	Fetch low-order operand address
		4	1	Fetch operand
		5	0	Store operand
JMP	3	1	1	Fetch op code
		2	1	Fetch high-order jump address
		3	1	Fetch low-order jump address

Table 7.3 Summary of cycle by cycle operations of the instructions.

necessary to introduce an additional type of machine cycle in which neither read nor write operations are involved. In that case, an additional control line is also required to distinguish it from the other two types of machine cycles. This, in fact, is the role played by the valid memory address (VMA) line in MC6800. The cycle-by-cycle operations of the MC6800 instructions are very similar to those of our present set of instructions.

Each machine cycle, whether it be a read-machine cycle or a write-machine cycle, can be represented as a state diagram. Consider the operations involved in a typical read machine cycle.

1. At the beginning of the cycle, the address coming either from the PC or DC, depending on whether an instruction or an operand is to be fetched, is moved to the memory addressing register (MAR), located in the memory unit. This can be achieved by issuing one of the following control signals:

   $$MAR \leftarrow PC$$
   $$MAR \leftarrow DC$$

   In some processors, more than one operation may be required to move the entire address from PC or DC into MAR.
2. When the data is ready in MBR, the memory buffer register, indicated by a signal on the READY line, control signals are generated to move the data through the data bus into the destination register, which can either be the accumulator AC or the high-order part or low-order part of the data counter (DC), or the instruction register IR.
3. Any logic or arithmetic operations required are then executed, making use of the various components in the ALU.

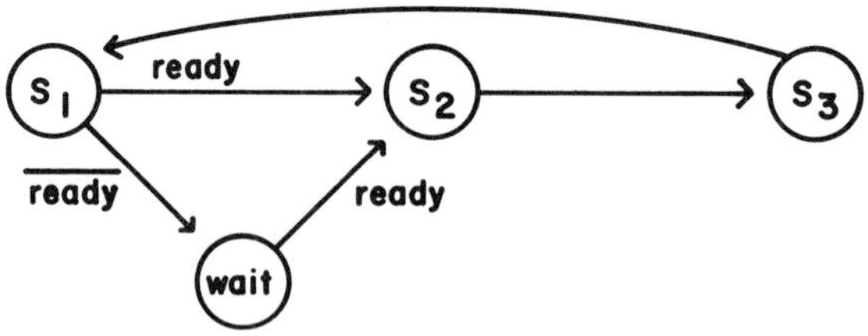

Figure 7.11 State diagram for machine cycle

State	X_1	X_2
S_1	0	0
S_2	0	1
S_3	1	0
Wait	1	1

Table 7.4 State variable codings.

We can draw a state diagram to represent a typical read-machine cycle following the construction procedure given in Section 4.1. This is shown in Fig. 7.11. During state S_1, the address is transmitted to the MAR. A wait state is included to allow for asynchronous memory accessing. During state S_2, the data available in the MBR is transferred to its destination register in the CPU. Any required ALU operation is carried out during state S_3. The same state diagram used to represent a read machine cycle can also be used to represent a write-machine cycle. In this case, during state S_2, the data in AC to be stored is moved to MBR where it is written into the location address specified in MAR. The write signal can be assumed to be generated during state S_3 although this is not necessary.

We thus have a state diagram which can be used to implement a machine cycle, whether it be a read or a write cycle. Since there are a total of four distinct states, including the wait state, two state variables X_1 and X_2 are required to identify the states and the coding chosen is shown in Table 7.4.

We are now interested in determining what control signals will be required in each state to achieve the execution of each of the instructions. We shall first examine each instruction and then collect all the results at the end. For brevity and clarity, direct gatings rather than busing notations are used. The reader should have no difficulty in modifying the design to use busing techniques. In addition, PC is assumed to be automatically incremented by one each time it is referenced.

LDI This instruction requires two machine cycle to execute. The following are the control signals which are required during each of the two machine cycles in each of the states (the WAIT state is omitted for obvious reasons).

Machine Cycle 1

S_1 MAR ← PC

S_2 IR ← MBR

S_3 nil

Machine Cycle 2

S_1: MAR ← PC
S_2: AC ← MBR
S_3: nil

SLL

Machine Cycle 1

S_1: MAR ← PC
S_2: IR ← MBR
S_3: nil

Machine Cycle 2

S_1: nil
S_2: nil
S_3: C,AC(0:7) ← AC(0:7),0

LDA

Machine Cycle 1

S_1: MAR ← PC
S_2: IR ← MBR
S_3: nil

Machine Cycle 2

S_1: MAR ← PC
S_2: DC(0:7) ← MBR
S_3 nil

Machine Cycle 3

S_1: MAR ← PC

S_2: DC(8:15) ← MBR
S_3: nil

Machine Cycle 4
S_1: MAR ← DC
S_2: AC ← MBR
S_3: nil

The sets of control signals that are required for the ADD, STA, ISZ, and JMP instructions can similarly be obtained and will be left as an exercise to the reader. All the control signals are summarized in Table 7.5.

Each of the instructions require from two to five machine cycles for complete execution. This sequencing of machine cycles can be done through the use of a counter having three output lines, C_1, C_2, and C_3 to identify the specific machine cycle of the CU.

We now summarize the information needed to generate a set of control signals during a particular phase in the execution of an instruction:

1. The particular machine cycle within the instruction, identified by control lines C_1, C_2, and C_3.
2. The type of machine cycle, indicated by the read/write (R/W) line, which informs the memory whether a read or write operation is to be performed.
3. The particular state within a machine cycle, which is identified by control lines X_1 and X_2. The WAIT state is entered and exited by detection of the value on the memory ready line coming from the memory unit.
4. The correct control signals within each state which are generated through the decoding of the instruction operation code together with information coming from certain parts of the CPU like the status register.

We have thus far obtained nearly all the necessary information needed to construct a sequential circuit for the control unit. Before the circuit can actually be specified, it is of course necessary to assign 8-bit operation codes to the various instructions in concern. This and the final complete design will be left as an exercise to the interested readers (see Problem

Machine cycle	State	LDI	LDA	SLL	ADD	STA	ISZ	JMP
		Instruction						
1	1	MAR←PC	MAR←PC	MAR←PC	MAR←PC	MAR←PC	MAR←PC	MAR←PC
	2	IR←MBR	IR←MBR	IR←MBR	IR←MBR	IR←MBR	IR←MBR	IR←MBR
	3	nil	nil	nil	nil	nil	nil	nil
2	1	MAR←PC	MAR←PC	nil	MAR←PC	MAR←PC	MAR←PC	MAR←PC
	2	AC←MBR	DC(0:7)← MBR	nil	DC(0:7)← MBR	DC(0:7)← MBR	DC(0:7)← MBR	DC(0:7)← MBR
	3	nil	nil	C.AC(0:7)← AC(0:7),0	nil	nil	nil	nil
3	1		MAR←PC		MAR←PC	MAR←PC	MAR←PC	MAR←PC
	2		DC(8:15)← MBR		DC(8:15)← MBR	DC(8:15)← MBR	DC(8:15)← MBR	DC(8:15)← MBR
	3		nil		nil	nil	nil	PC←DC
4	1		MAR←DC		MAR←DC	MAR←DC	MAR←DC	
	2		AC←MBR		buffer←MBR	MBR←AC	AC←MBR	
	3		nil		AC←ADD (AC,buffer)	nil	AC←ADD (AC,1)	
5	1						MAR←DC	
	2						MBR←AC	
	3						PC←ADD (PC,3Z)†	

Table 7.5 Control signals for the instruction subset.
† PC is incremented by 3 if the zero status-bit Z is 1.

7.3). The result of such a design is commonly known as a *hardwired* control unit.

Obviously the above described CU design example is oversimplified in several respects, a few of them being the following:

1. Only a subset of any typical instruction set is included. The design for a complete instruction set of reasonable capability, though much more complex, can be achieved by the same procedure.
2. As indicated at the beginning of this section, direct gating without the use of a bus has been employed for brevity as well as clarity. Adjusting the above design arguments to take care of busing should be straightforward.
3. No interrupt capability or other more sophisticated operating capabilities have been incorporated in the above design example. Some of these topics will be taken up in the next chapter.

Another approach which can be used to produce a hardwired control unit can be found in [1]. In this approach, the control unit is implemented through the direct use of delay elements. One advantage of this approach is that it becomes very convenient to first of all write down the entire design of the control unit in the form of a program using an algorithmic language called AHPL which is essentially a subset of APL tailored to the description of digital hardware. The actual logic circuit can then be obtained from the program by a translation process which can usually be done by a translator program. This approach thus lends itself readily to the automatic design of a CU with the aid of a computer system. One of the disadvantages of this approach is that in a complex system, an exceedingly large number of delay elements or equivalent components has to be used resulting in a very low-speed CU.

We shall now consider the design of a microprogrammed control unit which is becoming more and more common, especially in the design of microprocessors.

7.4 MICROPROGRAMMED CONTROL

The basic concepts of microprogramming were introduced by Wilkes [6] with the original primary objective of providing a systematic alternative to the somewhat ad hoc procedure usually used for the design of a hardwired control unit. This original objective still carries over to the present stage of computer development in some respects. Other more important advantages of microprogramming have created its immense popularity.

One of the more important advantages of microprogramming is the flexibility it affords in the choice of an instruction set which is of paramount importance in computer systems for specialized applications, such as the emulation of other computer systems.

The basic organization of a computer system with a microprogrammed control unit is shown in Fig. 7.12. The instructions (also known as *macroinstructions*) are stored in main memory and are accessed through the memory-address-register (MAR) and the memory-buffer-register (MBR). Each of the instructions is decoded in the local memory store giving rise to a sequence of one or more *microinstructions* which are obtained from the control store through the control address register (CAR) and the control buffer register (CBR). Each of the microinstructions in turn is responsible for the generation of one or more very primitive control signals. Note the functional resemblance between the register pair MAR and MBR to the pair CAR and CBR. The former pair is used to access macroinstructions while the latter pair is used to access microinstructions. There is also a microinstruction sequencing circuit similar in structure to an

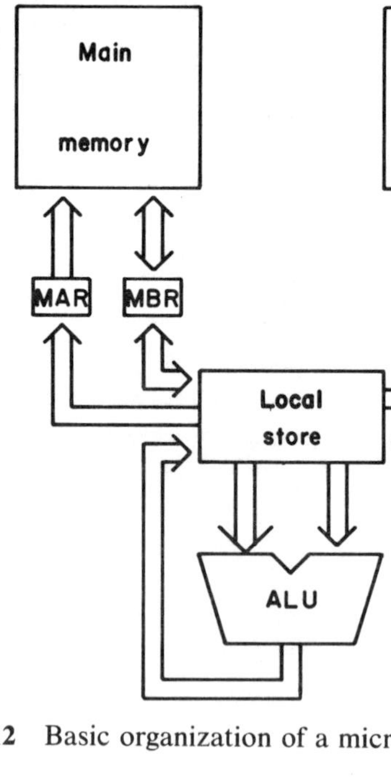

Figure 7.12 Basic organization of a microprogrammed computer

instruction sequencing circuit which makes use of a program counter which we shall denote as CC.

The control store is usually of ROM type memory since it is frequently desirable that microinstructions be stored as unalterable information. However, we shall shortly indicate the availability of microprogrammable computers in which the organization of the microinstructions can be altered.

Each of the one or more operations invoked by the control signals generated through the execution of a microinstruction is known as a microoperation. These are very primitive (simple) in nature, involving in most cases simply the control of gating logic for an ALU operation or the transmission of data to or from a bus. The corresponding sequence of microinstructions used to implement a microinstruction is known as a *microprogram* or *microroutine*. It is possible to have a certain amount of overlap between two or more microprograms implementing different macroinstructions.

A microprogrammed control unit is said to be microprogrammable if the organization of the control store in which the microprograms are stored can be modified by the user to produce tailored macroinstructions. If the control store is not intended to be alterable by the user, it is then not microprogrammable. In this case, the set of macroinstructions are fixed and are then comparable to a set of instructions found in a hardwired control unit. Most of the presently available microprogrammed processors are not user microprogrammable, except for a limited number of bit-sliced microprocessors in which the ALU has been organized in units or slices of certain widths (mostly 2-bit and 4-bit). (See Section 7.5 for a microprogrammable bit-sliced microprocessor.) These permit more flexibility in the organization of the CPU to better meet specific operational requirements, usually in terms of speed.

There are essentially two basic approaches to design a microprogrammed control unit, *horizontal microprogramming* and *vertical microprogramming*. In the former approach, a single bit in a microinstruction word is used to control a single microoperation so that if the system requires n distinct microoperations, then a microinstruction wordlength of n bits is required. Fewer bits would suffice but this approach has the advantage that as many microoperations as needed can be initiated simultaneously permitting fastest operation. Most of the microoperations are mutually exclusive and are never invoked simultaneously in which case it is possible to divide them into groups and to use a number of bits (a field) to code a set of mutually exclusive microinstructions. A decoder is used to select the particular operation to be invoked. This coding and decoding if carried to an extreme (only one field) result in vertical microprogramming in which only one microoperation is invoked at a time. Most processors employ a combination of the two approaches. A majority of the new microprocessors have the fields divided in such a way as to provide the added advantage of permitting the user to more easily re-microprogram the control unit to suit a particular operational requirement.

Consider the block diagram of a typical microprogrammed control unit as shown in Fig. 7.13. The reader will immediately realize the similarity between the microprogrammed control unit and a CPU with main memory. The ROM together with CAR and CBR where the microprograms are stored, has a counterpart in the main memory with MAR and MBR. The microinstruction register (CIR) is a counterpart of the instruction register (IR). The microsequencer with the microinstruction counter is the counterpart of the control unit C with the program counter PC. Status information needed for conditional branchings must be supplied, and this is usually done by a small set of data registers within the microsequencer block.

The micro-sequencer is invoked by decoding a macroinstruction in IR. Once invoked, the micro-sequencer causes the following sequence of events:

1. The first microinstruction in the microprogram implementing the macroinstruction is fetched from ROM.
2. The fetched microinstruction is decoded to produce corresponding control signals.
3. Depending on the condition, one of two possibilities occur next: The next microinstruction in sequence is fetched, a branch is executed by fetching another microinstruction not in sequence (but in the same microprogram). The system than repeats (ii) or exits from the instruction interpretation cycle when the end of the microprogram is encountered.

As previously indicated, each of the microoperations generate very primitive control signals, like the moving of the information contained in a set of latches to the data bus or the setting of a flip-flop. These usually require only one clock cycle to complete although in some of the more sophisticated microprogrammed control units, some microoperations may

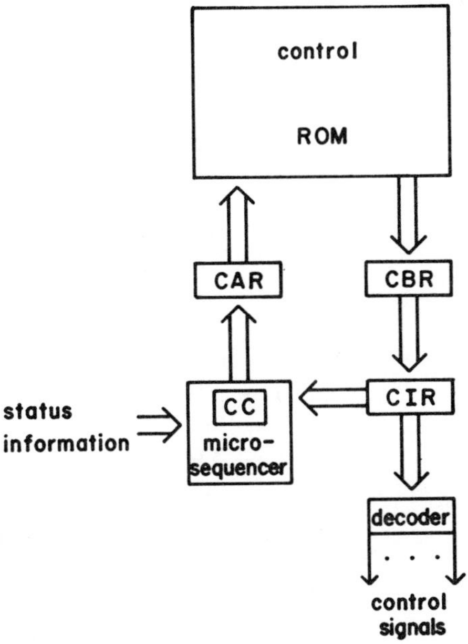

Figure 7.13 Block diagram of a control unit

take more than one clock cycle, thus requiring some kind of overlapping capability.

The format of a microinstruction can obviously vary significantly from design to design and depends on factors such as the set of microoperations chosen, the trade-off between microinstruction word length and speed of operation (horizontal vs. vertical), and the accepted degree of complexity of the microinstruction decoding circuits.

We will first determine a set of microoperations that are needed to implement the subset of instructions given in Table 7.1. Note that the set is not unique since it depends on the exact prescribed register organizations and data paths. The set that is chosen is shown in Table 7.6 in which each of the microoperations has been identified by an associated number. One microoperation which is unique in nature is microoperation 24 (the end microoperation) which is used to inform the microsequencer of the end of a microprogram. As a result, a new instruction interpretation cycle is started.

The first seven microoperations are used to fetch the instruction from memory into IR and are therefore required in every microprogram. The following microprogram implements the operation of the instruction ADD (three bytes long):

$$
\begin{aligned}
\text{ABUS} &\leftarrow \text{PC}\\
\text{MAR} &\leftarrow \text{ABUS}\\
\text{R/W} &\leftarrow 1\\
\text{PC} &\leftarrow \text{PC} + 1\\
\text{DBUS} &\leftarrow \text{MBR}\\
\text{IR} &\leftarrow \text{DBUS}\\
\text{ABUS} &\leftarrow \text{PC}\\
\text{MAR} &\leftarrow \text{ABUS}\\
\text{PC} &\leftarrow \text{PC} + 1\\
\text{DBUS} &\leftarrow \text{MBR}\\
\text{DC}(0:7) &\leftarrow \text{DBUS}\\
\text{ABUS} &\leftarrow \text{PC}\\
\text{MAR} &\leftarrow \text{ABUS}\\
\text{PC} &\leftarrow \text{PC} + 1\\
\text{DBUS} &\leftarrow \text{MBR}\\
\text{DC}(8:15) &\leftarrow \text{DBUS}\\
\text{ABUS} &\leftarrow \text{DC}\\
\text{MAR} &\leftarrow \text{ABUS}\\
\text{DBUS} &\leftarrow \text{MBR}\\
\text{BUFFER} &\leftarrow \text{DBUS}\\
\text{ADDER} &\leftarrow \text{BUFFER, AC}\\
\text{AC} &\leftarrow \text{ADDER}\\
&\text{end}
\end{aligned}
$$

The opcode is first fetched from memory and placed in the IR. (Note that the content of the PC is incremented after each reference to the PC). Then the two bytes of information representing the address of the operand are fetched and placed in the data counter DC. Finally the value of the operand is fetched and then the adder in the ALU is used to perform the addition. The result replaces the old value in AC.

The following is a microprogram for the instruction LDI (2 bytes long):

$$
\begin{aligned}
\text{ABUS} &\leftarrow \text{PC}\\
\text{MAR} &\leftarrow \text{ABUS}\\
\text{R/W} &\leftarrow 1\\
\text{PC} &\leftarrow \text{PC} + 1\\
\text{DBUS} &\leftarrow \text{MBR}\\
\text{IR} &\leftarrow \text{DBUS}\\
\text{ABUS} &\leftarrow \text{PC}\\
\text{MAR} &\leftarrow \text{ABUS}\\
\text{PC} &\leftarrow \text{PC} + 1\\
\text{DBUS} &\leftarrow \text{MBR}\\
\text{AC} &\leftarrow \text{DBUS}\\
&\text{end}
\end{aligned}
$$

Microoperation number	Microoperation
1	ABUS ← PC
2	MAR ← ABUS
3	R/W ← 0
4	R/W ← 1
5	PC ← PC + 1
6	DBUS ← MBR
7	IR ← DBUS
8	AC ← DBUS
9	SHIFTER ← c,AC
10	c,AC ← SHIFTER
11	DC(0:7) ← DBUS
12	DC(8:15) ← DBUS
13	ABUS ← DC
14	BUFFER ← DBUS
15	ADDER ← BUFFER,AC
16	AC ← ADDER
17	DBUS ← AC
18	MBR ← DBUS
19	ADDER ← BUFFER,1
20	increment PC by 3 if Z = 1
21	PC(0:7) ← DBUS
22	PC(8.15) ← DBUS
23	PC ← ABUS
24	end

Table 7.6 A list of chosen microoperations.

Note that for simplicity in the above microprograms, and in the set of microoperations listed in Table 7.6, no consideration has been given to the possibility of waiting for slow memory units. In the hardwired design described in the last section this problem is handled by the introduction of a WAIT state. Here we can simply incorporate a microoperation which continuously tests the status of a ready line coming from the memory unit, and exits from this testing loop only if the signal on the line is a 1 indicating the condition that the desired value is ready in MBR.

The above set of microoperations can be implemented using a 24-bit word format. Let bit i (numbered from left to right) be used to code microoperation number $i + 1$ for $i = 0$ to 23. Then the following microcode (i.e. a word used to encode the microoperations)

$$\underbrace{1000 \ldots\ldots\ldots\ldots.\ 00}_{\text{24-bits}}$$

represents the microinstruction

ABUS←PC

It is usually possible to execute more than one microoperation simultaneously. For example, the two microoperations

ABUS←PC
R/W←1

can be done simultaneously since they affect different registers and use different logic circuits. Therefore we can combine their separate codes to form the following microcode which would cause both microoperations to be executed simultaneously

$$\underbrace{10010 \ldots\ldots\ldots\ldots.\ 0}_{\text{24-bits}}$$

By combining all nonconflicting microoperations as much as possible in each microcode, the length of a microprogram in terms of the number of microcodes can be significantly reduced. This is actually the major advantage of using horizontal microprogramming–maximum speed can be achieved, since the number of clock cycles needed to execute a microprogram is normally directly related to the length of the microprogram. Thus the fetching of an instruction op code can be done by the following sequence of 4 microcodes:

10010000 0
01001000 0
00000100 0
00000010 0

There are certain microoperations which form sets in which members are mutually exclusive of each other in the sense that if one microoperation in a set is executed then none of the other microoperations in the same set can be executed at the same time. For example, the two microoperations

$$R/W \leftarrow 0$$

$$R/W \leftarrow 1$$

could never be executed simultaneously. We can combine the members in each of the sets together by using a number of bits in the microcode word for the set and assigning codes to distinguish among the various microoperations in the set. The correct control signals are generated by making use of decoders. For example, suppose we have 5 mutually exclusive microoperations. Normally they would occupy 5 bits in a microcode word. Instead we can use 3 bits only to code these 5 microoperations at the expense of a decoding circuit as shown in Fig. 7.14. Note that to code 2 mutually exclusive microoperations, 2 bits are required, since the particular combination 00 has to be used to indicate the absence of both microoperations.

As a result of the above coding process, the word length of a microcode can be significantly reduced. This is very important since the number of distinct microoperations in a typical computer is a few hundred and the use of microcode word lengths of a few hundred is impractical. On the other hand, the use of very complex decoders is also undesirable and so in general, a compromise between the microcode word length and the decoder complexity is necessary.

As an exercise, the reader should design a microcode word format for the set of microoperations listed in Table 7.6 limiting the size of decoders used to those with at most 3 input lines.

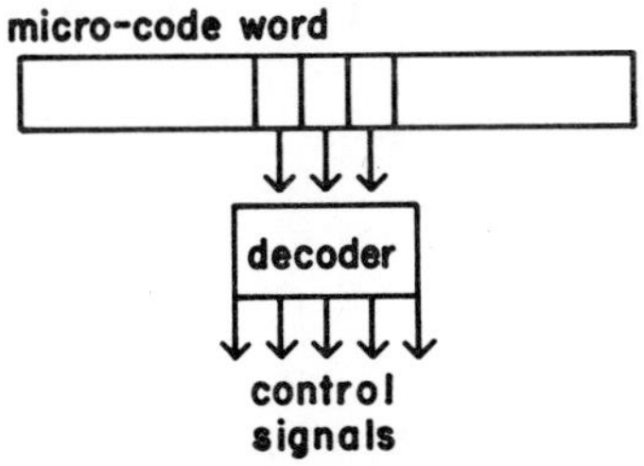

Figure 7.14 In this example, 3 bits are used to code 5 mutually exclusive micro-operations

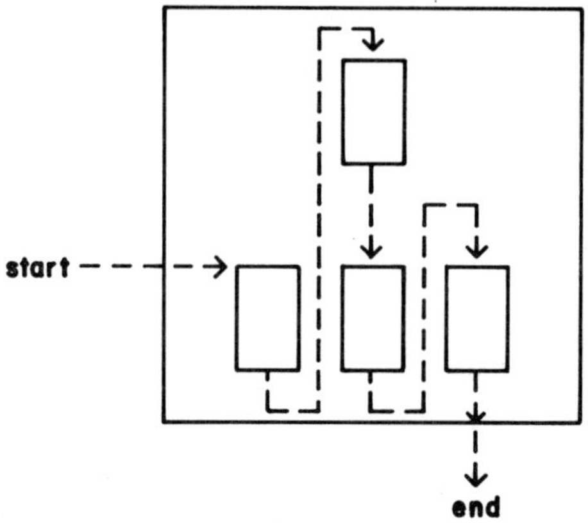

Figure 7.15 Branching within the control ROM. Each box represents a block of microcodes

If the microprograms in the control ROM can be altered by the user, the processor is said to be microprogrammable. In this case the user can actually design his or her own instruction set (the set of macroinstructions). Otherwise the processor is not microprogrammable. Most large-scale computers as well as minicomputers which are microprogrammed are not microprogrammable whereas a majority of the microprogrammed microprocessors are microprogrammable.

Now let us look at the role played by the microsequencer and see how it can be constructed. Two microprograms implementing two macroinstructions very often include blocks of identical microcodes. For example, each microprogram has at the beginning a set of microcodes used for fetching the instruction operation code. A large amount of the control ROM would be wasted if separate copies of the blocks, one for each of the instructions, are kept. In order to avoid this, logic is included in the microsequencer to permit picking the way around the control ROM. This is illustrated in Fig. 7.15. This allows some of the codes to be shared by a number of microprograms. The correct branching is achieved by looking at the particular instruction being interpreted, as well as one or more of the status flags. A list of the typical functions that should be included in a micro-sequencer is given in Fig. 7.16.

It is not difficult to see that the microsequencer with respect to microinstructions has essentially the same role as an instruction sequencer. However, the structure of a microsequencer is much simpler due to a number of factors which include the following:

1. The microinstructions are all of the same length while the instructions are not.
2. A much smaller number of registers are involved
3. The addressing space is considerably smaller
4. Timing considerations need not be involved especially when each microinstruction is assumed to take a fixed number of clock periods to be completed.

a. execute a contiguous sequence of microinstructions:

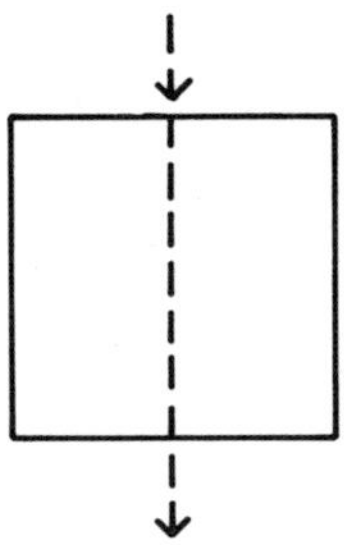

b. branch to another sequence of microinstructions:

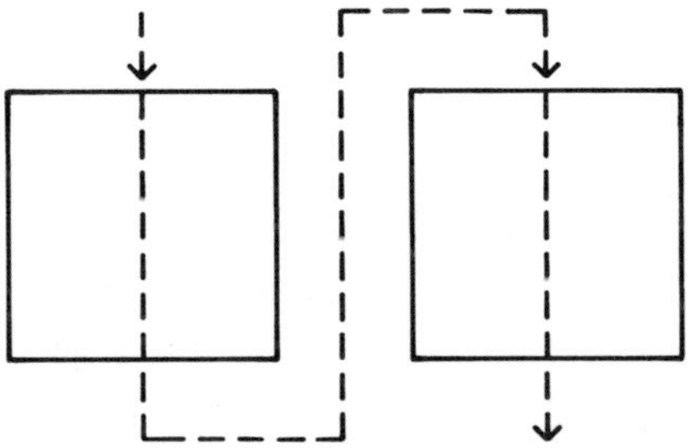

c. branch to and return from another sequence of microinstructions

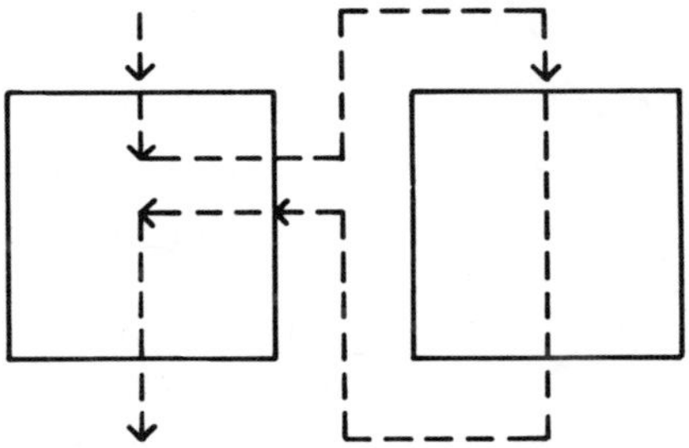

Figure 7.16 Functions needed in a microsequencer

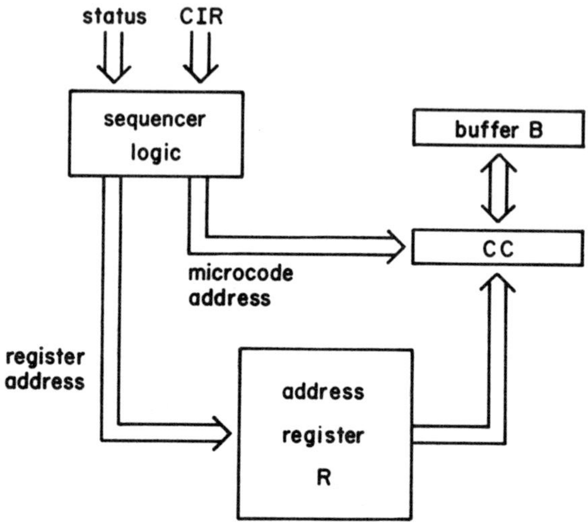

Figure 7.17 Block diagram of a microsequencer

5. In order to simplify control, only a small number of very frequently used common blocks of microcodes are shared by different microprograms.

A block diagram of a circuit which can be used to implement a microsequencer is shown in Fig. 7.17. CC is the microinstruction counter. Address register *R* is a small set of registers in which are permanently stored the addresses of the set of commonly used and shared blocks of microcodes. Buffer *B* is a register which is used to temporarily store the return address when a subroutine type of return is required. (See Fig. 7.16(c)). The techniques introduced in Chapters 2 and 3 can easily be adopted to design such a sequencer logic circuit.

7.5 AN EXAMPLE OF A MICROPROGRAMMABLE PROCESSOR

In this section, we briefly describe the operation of an existing microprogrammed control unit and its associated components, the Intel 3000 Series family of bipolar elements. We first present a block diagram of a typical system and then describe the various components of the system individually. The system is user microprogrammable and in addition allows for a high degree of flexibility in the form of architectural configuration based on the concept of bit slicing. As shown in Fig. 7.18, a typical control processing unit built around the Intel 3000 family is made up of three components: the control memory (CM) for storing the collection of microprograms implementing the macroinstruction set, the micro-control unit (MCU) as the microsequencer, and the central processing element (CPE) array for performing various arithmetic and logic functions. A CPE is a 2-bit slice of ALU so that an array of the appropriate number of CPEs can be connected to form an ALU of the desired length; e.g., an array of 8 CPEs results in an ALU which is 16 bits long.

The format of a microinstruction is variable and contains certain fields which are optional and user-defined as shown in Fig. 7.19 to allow for the aforementioned architectural flexibility. The microinstruction word contains three fields which are standard for all configurations: (i) seven bits are used to select and control the functions performed in the CPE array; (ii) four bits are used to select and control the use of the carry and zero status flag flip-flops residing in MCU; (iii) seven bits are used to supply information to the MCU to determine the address of the next microinstruction to be executed. The rest of the fields are user-defined and are used for controlling optional features. An example of this is the mask field which is used to provide mask bits to the CPE array to increase

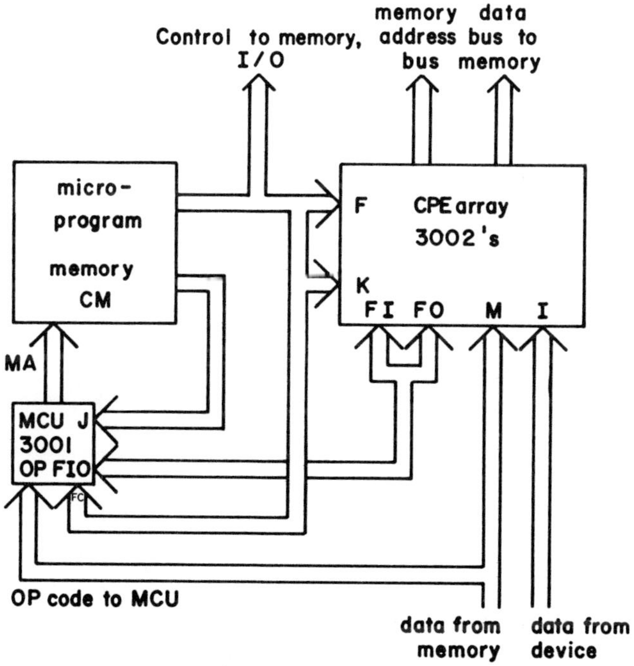

Figure 7.18 Block diagram of a typical system

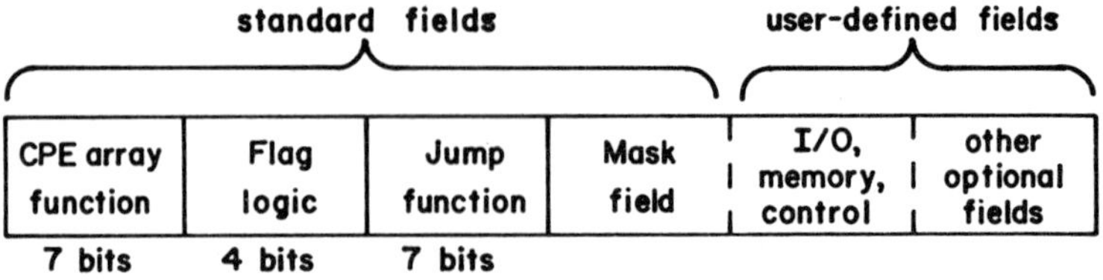

Figure 7.19 Microinstruction format

the versatility of the CPEs. The length of each of these user-defined fields is also variable. This means that the wordlength of a microinstruction is dependent on the particular configuration used in the system. The collection of microinstructions, in the form of microprograms each of which is being used to implement a macroinstruction, is stored in the control memory unit (CM). When a microinstruction in CM is addressed by the MCU, the various fields in the microinstruction are read and transmitted to their respective destinations to perform their associated control functions.

The internal function to be performed in the CPE array is selected by the 7 input bits at F from the control memory CM with certain masking of operand bits being controlled by the input bits at K (also from CM). Data from main memory and I/O devices are entered at the terminals M and I and there are also terminals going to the memory-address bus and data bus. There is, in addition, one terminal each for the carry input, carry output, shift right output, and shift left input (FI and FO) used to transmit status bits and shift bits to and from the MCU.

The MCU acts primarily as the microsequencer. The operation code of the macroinstruction comes in at OP from main memory and is decoded to the starting address of its corresponding microprogram. This address, sent through MA, is used to fetch from CM the first microinstruction. Within a microprogram, the next address is formed from a number of sources of information: the MCUs current microprogram address, the jump function field bits at J, the operation code, and the C and Z status flags in MCU. The other major control task for which the MCU is responsible involves keeping track of the carry and shift inputs and carry and shift outputs of the CPE array by means of the flag logic control bits at FC.

The preceding is only a brief description of the structure of a typical system based on the Intel 3000 Series which incorporates many interesting and innovative features in microprogrammed control unit design. The system is user microprogrammable in that the user can choose his or her own macroinstruction set and to a certain extent, his or her own system configuration. Then all that is needed is to write the appropriate microprograms.

PROBLEMS

7.1 Divide 11 by 3 using the restoring division method. Give all intermediate and final results. Assume that the dividend is in AC-MQ and the divisor is in register DV. The quotient should be in MQ and the remainder in AC. Use 8-bit registers.

7.2 Repeat Problem 7.1 using the nonrestoring division method.

7.3 Assume that a CPU is to be designed to execute only the instructions ADD and LDA (Table 7.5) which have been assigned the following operation codes:

ADD: 00001111 LDA: 00001110

Complete the design of the control unit as summarized at the end of Section 7.3. Include a complete logic diagram of the circuit obtained. You are allowed to use standard components, like decoders, etc.

7.4 Write a microprogram for the macroinstruction JMP as defined in this chapter in terms of the microoperations in Table 7.6.

7.5 Write a program in MC6800 machine language to multiply two 3-bit positive integers.

7.6 Convert the program in Problem 7.5 into a subroutine and write a main program to multiply any two integers expressed in two's complement form, for values from −8 to +7 making use of the resulting subroutine.

7.7 Design a microinstruction word format for the set of microoperations given in Table 7.6. Assume that only decoders with at most 3 input lines are to be used in the generation of control signals (Fig. 7.14). The design will not be unique but try to minimize the length of the microinstruction word.

7.8 Based on the diagram in Fig. 7.17, draw a flowchart describing the operation of the sequencer logic.

REFERENCES

1. Hill, F. J. and Peterson, G. R. *Digital Systems Hardware Organization and Design*. New York: John Wiley & Sons, 1973.
2. Husson, S. S. *Microprogramming Principles and Practices*. Englewood Cliffs, N.J.: Prentice-Hall, 1970.
3. Motorola, Inc. *M6800 Microcomputer Systems Design Data*. 1976.

4. Osborne, A. *Introduction to Microcomputers vol. I, Basic Concepts*. Berkeley, CA: Adam Osborne and Associates, Inc., 1976.

5. Intel Corporation. *Series 3000 Reference Manual*. 1976.

6. Wilkes, M. V. The Best Way to Design an Automatic Calculating Machine. *Report of Manchester University Computer Inaugural Conference*. July 1951, pp. 16–18.

7. Zaks, R. *Microprocessors from Chips to Systems*. Berkeley, CA: Sybex, 1977.

Chapter 8

INTERFACING: I/O TRANSFER AND INTERRUPT DESIGN

We have thus far concentrated on the design of the various components of a computer system on the assumption that these components will ultimately be interconnected to form the system desired. Interconnection of the internal components of the system like main memory, RAM or ROM, and the CPU can easily be achieved through the use of the system bus. Information, either data or instructions or various control signals, is transferred between CPU and main memory through the system bus, either synchronously or asynchronously, directly controlled by the control unit CU. This situation has been discussed both in Chapter 5 and also in Chapter 7, where the design of the CPU is pursued.

On the other hand, interfacing external I/O components to the computer system is not as straightforward mainly due to the fact that I/O devices exhibit a diversity of operational characteristics such as signal types, coding, as well as speeds as discussed in Chapter 1. This aspect of design is generally known as interfacing beyond the computer system and of course deals with the transfer of information to and from the computer system and external devices. Different applications call for different types of interfacing methods and implementations. Some of the resulting interface units can perform only very primitive functions such as the transformation of digital values to appropriate voltage levels to light indicators or to activate switches, while some others may have capabilities comparable to those of mini-computers. A block diagram of interfacing is shown in Fig. 8.1. Note that each interface can be connected to more than one I/O device.

The transfer of information between the computer and external devices basically consists of the following operations:

1. checking the device for readiness,
2. initiation of transfer when the device is ready,

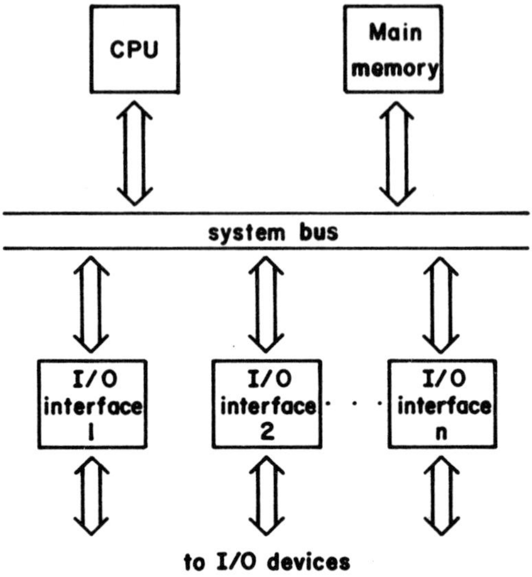

Figure 8.1 Block diagram of interfacing

3. the actual transfer of information, and
4. conclusion of transfer.

The details involved in each of the above operations vary and are dependent on the particular operational environment. Depending on the desired degree of intelligence to be incorporated into the interface, there is one of three basic approaches which can be used to design an interface to accomplish a data transfer. These three design approaches are essentially based on three methods that can be used to transfer data to and from the computer system and the external devices. The three data transfer methods are:

1. programmed transfer,
2. interrupt transfer,
3. direct memory access (DMA) transfer.

These three transfer methods form the basis of the material in this chapter and we shall look at each of them and their associated interface circuits in detail in the next few sections. The programmed transfer method is probably the most instructive one as far as understanding the flow of information involved in an I/O operation is concerned and so it will be discussed first.

Note that an I/O device can itself be another computer, thus resulting in a multiple-computer system or a computer network. Some of the more commonly used I/O devices include switches, keyboard, LED displays, paper-tape reader, card reader, line printer, magnetic discs, drums, CRT displays, magnetic types, and teletypewriter. A number of these devices operate on parallel data, typically one byte at a time, while others operate on serial-bit data, one bit at a time, requiring parallel-to-serial and serial-to parallel transformation of the input and output information. Physically remote I/O devices must be connected through communication links. This topic will be discussed further in Section 8.5.

8.1 PROGRAMMED TRANSFER

In programmed transfer, all input/output operations are completely initiated and controlled by the CPU through the execution of a program. Thus, whenever a data transfer is required, a special program is executed to accomplish the data transfer operations. The CPU is responsible for keeping track of all the details that are needed to perform the operations. As a result, the amount of logic required to implement an interface based on this type of data transfer is minimal. Associated with each I/O device or each set of I/O devices is one or more I/O buffers connected to the data bus as shown in Fig. 8.2. The content in an I/O buffer is gated into

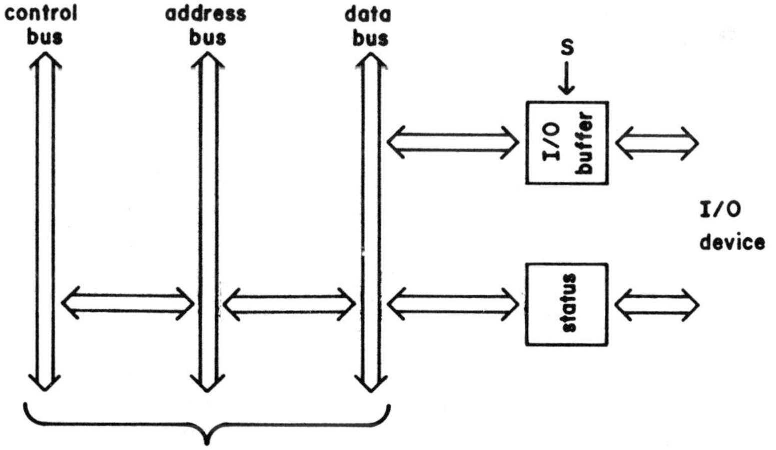

Figure 8.2 I/O buffer connected to a data bus for programmed transfer

or from the data bus by the application of a select signal at terminal S which is a signal resulting from the decoding of an I/O device address transmitted through the set of address lines in the address bus or a subset of these lines if only part of the address space is used.

There are essentially two ways to look at an I/O device, or equivalently, the associated I/O buffers. One is to regard it as nothing but a set of storage locations just like ordinary storage locations in main memory. In this way, no special I/O instructions are required. Part of the addressing space is reserved for addressing the various I/O buffers at the expense, naturally, of the directly addressable main memory space. One of the address lines, e.g., that corresponding to the highest order bit, can be used to distinguish between ordinary main memory addresses and I/O buffer locations. This approach is taken in a number of computer systems, notably the DEC PDP-11 and some microprocessors like the MC6800. This is seldom done, however, in large-scale computers where maximum direct addressability of the main memory space is always desired.

The second way is to have a special set of I/O instructions to take care of data transfers coming from and going to I/O buffers. Let us use the mnemonic code IOT (I/O transfer) for the I/O instructions. A possible format for a set of I/O instructions is that shown below:

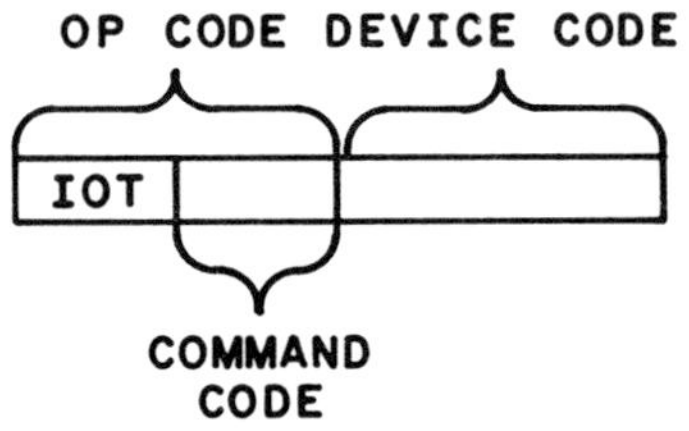

where the command code is used to identify the required operation, e.g. a read or write operation, and the device code is a preassigned number used to identify each of the I/O buffers in the system. If the accumulator or a particular register in a multiple-register CPU is the implied destination or source of the I/O transfer, then no other addressing information need be supplied. Otherwise, additional bytes of information needed to reference a particular memory location has to be included as part of the instruction.

To achieve synchronization with external devices, additional instructions and logic in the interfaces are generally required. One example is an instruction that is usually needed to test a status flag of an I/O buffer, where the status flag is a flip-flop used to indicate, for example, whether the I/O device is or is not ready to transmit or receive data. In larger

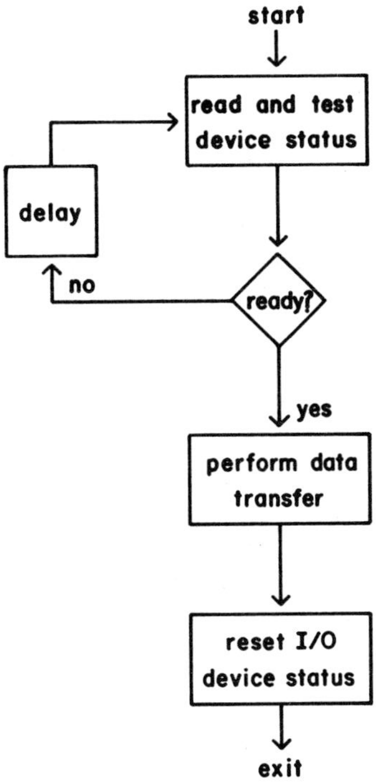

Figure 8.3 Flowchart for programmed data transfer

computers, this status information is usually transmitted through a separate set of control lines whereas in microcomputers, the data lines, or a subset of these data lines, are often used for this purpose.

A flowchart depicting the necessary operations required to achieve a programmed transfer between the CPU and an I/O device is shown in Fig. 8.3. Note the presence in the flowchart of a programmed loop used to detect the readiness of the I/O device for transmission of data through repeated reading and testing of the I/O device's status bits.

8.2 INTERRUPT TRANSFER

It is not difficult to realize the disadvantage of using the programmed transfer method in situations where the CPU has to wait a long time for the I/O buffer in concern to be ready for either reception or transmission of data–the CPU, while waiting, is unable to perform other operations. As a result, the level of performance of the entire system can be severely degraded. In situations like this, it is desirable to have the CPU informed of the readiness of the I/O buffer by the interface. The CPU will normally be doing some other useful work and as soon as it is informed of the readiness of the I/O buffer it temporarily suspends the work it is currently doing, and then switches over to take care of the I/O transfer. This kind of I/O transfer is known as transfer by interrupt. The I/O interface sends an interrupt request to the CPU whenever it is ready to do an I/O transfer. The CPU is interrupted, suspends its normal operations, and performs certain special operations to take care of the interrupt request. This means that the I/O interface has to have a certain amount of intelligence, more than that required in the case of programmed transfers. It has to be able to send interrupt requests to the CPU at appropriate times to alert the CPU of certain operations that need to be performed.

At the beginning of each instruction cycle, the CPU checks to see if there is any interrupt request issued by the I/O devices. If there is none, then the normal instruction execution operations are performed. On the other hand, the presence of an interrupt request forces the CPU to store the status of the presently executing program in a set of fixed locations in the CPU or, more often, in main memory and control is then passed on to one of a set of special routines, also stored in memory. The status of an executing program usually includes the contents of the PC, the AC, and various status flags sufficient to permit resumption of program execution later on. This collection of information is generally known as the program status and is stored in a set of locations in memory generally shown as the program status word (PSW). For example, the 64-bit PSW implemented in IBM 360 machines has a format as shown in Fig. 8.4. The reader is referred to the IBM assembler language reference manuals for a detailed description on the various fields in the PSW and of how interrupts are handled. Two of these fields, holding the instruction address coming from the PC and the interruption code, are especially basic in the

system mask	key	AMWP	interruption code	I C L	C C	program mask	instruction address

AMWP: Current status of computer
ILC: Instruction Length Code
CC: Condition Code

Figure 8.4 PSW in IBM 360

implementation of interrupts. The instruction address is used to return control to the point where execution is to be resumed in the interrupted program as soon as all interrupts have been handled. The interruption code indicates what caused an interrupt so that the appropriate interrupt handling routine can be invoked. We shall look at what an interrupt-handling routine actually does in a moment.

The presence of one or more pending interrupts is indicated by a signal on an interrupt line, IREQ, included in the set of control lines. The CPU acknowledges reception of an interrupt by sending a signal through the IACK control line. This technique of synchronization between the CPU and I/O interfaces is known as hand-shaking. At the same time, the CPU generally disables a special interrupt flip-flop to prevent itself from being interrupted again before the present interrupt has been serviced. If this is done we have a *single-level interrupt system.* We shall see later how a multiple-level interrupt system can be implemented. Next the value of the PC is stored in a fixed location, and the starting address of an interrupt service routine residing in fixed memory is fetched and placed in PC and the execution of the routine can now be started.

The service routine performs the following events:

1. The content of the AC and various important status information in the program status word is stored.
2. It is determined which interface caused the interrupt–this can be done by either polling the interfaces by means of a special instruction or having the interrupting interface send in an identifying address (see Section 8.4).
3. The required data transfer is initiated.
4. The interrupting interface's interrupt request flip-flop is reset to indicate the completion of the service.
5. The interrupt flip-flop is enabled, typically by clearing a flag bit, to permit servicing of another interrupt request.
6. The interrupted program's status is restored so as to be able to resume its execution.

The interface therefore must be able to do the following tasks:

1. send an interrupt request through IREQ,
2. interpret the acknowledge signal on IACK,

3. transmit its identifying address to the CPU usually through the address bus lines if necessary.

The design of an interface having these capabilities will be described in more detail in Section 8.4 where we shall be discussing interfacing in a microprocessor-based system.

When there is more than one interface capable of making an interrupt request at the same time, it is necessary to employ a priority circuit like the priority encoder in Problem 2.20 to serve the requests one at a time. The interfaces are assigned fixed priorities. Another approach that can be taken is known as daisy chaining in which the interfaces are chained together and assigned priorities depending on their relative positions on the chain, with the head of the chain having the highest priority and the end of the chain having the lowest. The IACK signal is sent from the head and down this chain until it reaches and is interpreted by an interface that has issued an interrupt request. The obvious advantage of using daisy chaining is that the number of interfaces that can be connected to the system using only a single IREQ is not limited. (See Fig. 8.5).

Another situation arises when we want to have more than one level of interrupt. This means that we want to be able to interrupt a currently executing interrupt-handling routine. This can be of significant benefit since in practice certain data transfers to or from certain I/O devices if delayed for too long can be lost or can drastically deteriorate the overall performance of the system. These data transfers should be serviced ahead of others. Different levels of priorities of interrupts can be achieved by having more than one IREQ line with different preassigned priorities. Each IREQ has associated with it a different service routine. For this organization, a stack, implemented either by a set of special registers or

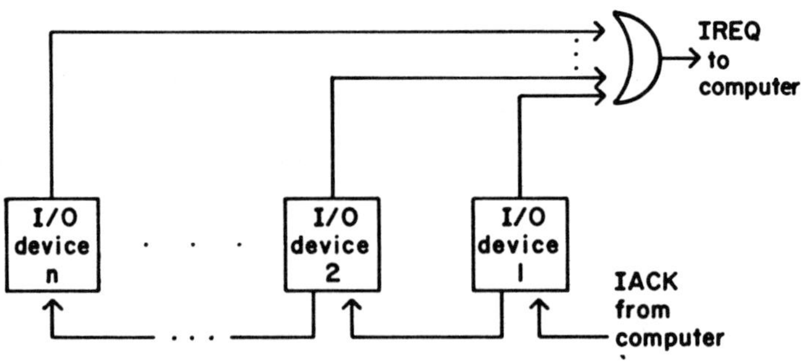

Figure 8.5 Daisy chaining of I/O devices

in main memory is generally needed to permit nesting of interrupt handling routine calls.

8.3 DIRECT MEMORY ACCESS

Transferring of data through the use of interrupts eliminates the situation where the CPU has to wait for the I/O interfaces to be ready for data transfer. However, the actual data transfers are still fully controlled by the CPU and during those periods, the CPU is incapable of doing anything else. A method which relieves the CPU of this duty is the *direct memory access* (DMA) method. In DMA, no software service routines are involved and all the data transfer operations, once initiated by the CPU, are done mostly or entirely by hardware, some located within the computer and some in the interface.

The DMA hardware in the interface usually contains the following set of registers:

1. an address register AR which is used to point to a location in main memory from or to which data is to be transferred. The original content in AR is loaded by the CPU and after that it is incremented by 1 automatically to point to the next location after each memory cycle (the time needed by the memory to complete one data transfer),
2. a register LR for storing the address of the last memory location where data transfer is to terminate.

Fig. 8.6 shows the block diagram of a DMA system illustrating the communication between the interface and the computer. The interface contains sufficient logic to make a DMA request to the computer. The heart of the DMA system resides in the DMA control unit which can either be an integral part of the computer or a component of the interface.

A DMA tranfer is initiated by a request made by an I/O device through the REQ control line and recognized by the CPU at the beginning of an instruction cycle. The CPU responds by sending to AR and LR the starting and final addresses of the set of consecutive locations to or from where the data are to be transferred followed by a signal on the ACK line. Once these initial steps have been performed, the CPU is free to resume its normal execution. However, each time an actual data transfer is to be made, the CPU activities are suspended for one memory cycle so as to permit the transfer to be done. This is known as *cycle stealing* and can

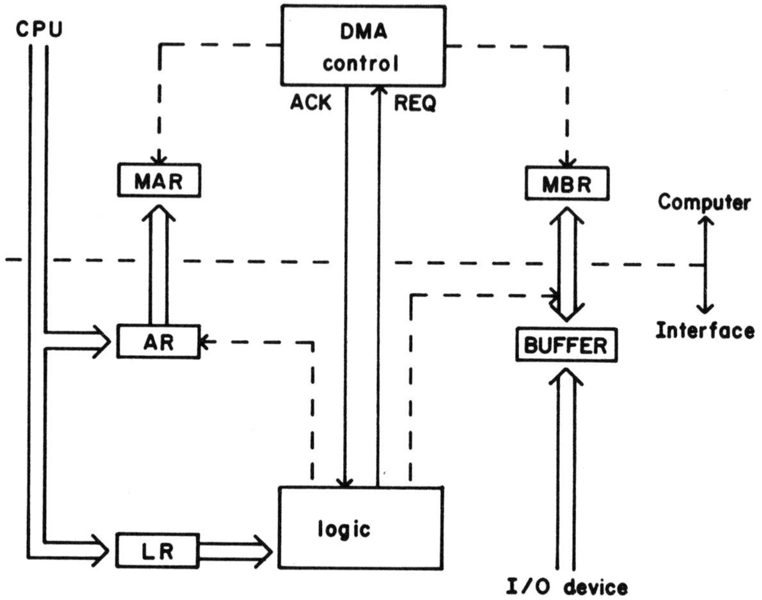

Figure 8.6 Block diagram of DMA systems

easily be achieved in a number of microprocessor systems by, for example, temporarily stopping the system clock for the duration of one memory cycle. Note that, here, the CPU does not have to store and subsequently restore program status information since the internal registers do not take part in the data transfer.

In the DMA system described above, the CPU is still held up for one memory cycle each time an actual data transfer is being made. In cases where the memory unit is capable of supporting concurrent multiple accesses with multiple MARs and MBRs, it is quite easy to further eliminate this need to steal a memory cycle from the CPU by providing a separate set of addresses, data, and control buses for the DMA system. The CPU and the DMA system can now operate almost independently except for times when conflicting memory accesses from the CPU and the DMA system are detected and have to be resolved; only the initiating steps must be done by the CPU.

Almost all large-scale computers have incorporated the DMA facility and an increasing number of minicomputer and microcomputer systems have this as an optional feature to increase their range of applications.

8.4 INTERFACING IN A MICROPROCESSOR-BASED SYSTEM

In this section, we consider a typical interface unit for a microprocessor-based computer system. This interface unit uses the interrupt method to transfer bytes of data to and from an I/O device. A block diagram of the unit, showing its three major functional components, the data bus buffers, the I/O registers, and the select and control circuit, is given in Fig. 8.7.

An expanded block diagram of the interface circuit is shown in Fig. 8.8. When a transfer operation is required by an I/O device, an interrupt request is sent from the select and control circuit to the CPU through an interrupt line IREQ. The CPU, on receiving this interrupt signal, sends an interrupt acknowledge signal back to the interface through the IACK line to start the interrupt service operations. The address assigned to this interface unit is used to identify this particular interface unit–bit A_{15} is used to distinguish between main memory addressing and I/O buffer addressing, A_{14} to A_1 are used to distinguish among the various interfaces connected to the CPU and A_0 is used to distinguish between the two I/O registers in the interface, the status register SR, and the data register DR.

The read/write line is used to control the direction of data flow. For reading, data is fetched either from DR, representing input data from the I/O device, or from SR, representing information concerning the status of the interface unit or the I/O device in concern. DR is one byte in length since 8 bits of information are treated as a unit. The length of SR is also one byte although, depending on the particular I/O device(s) in concern, some of the 8 bits may be unused. For writing, data from the CPU is now placed either in DR as output data to the I/O device or in SR as signals needed to control the various operations of the I/O device(s), like the rewinding of tapes, etc. If more than one I/O device is to be connected to the interface, multiplexors will be required to achieve sharing of the interface.

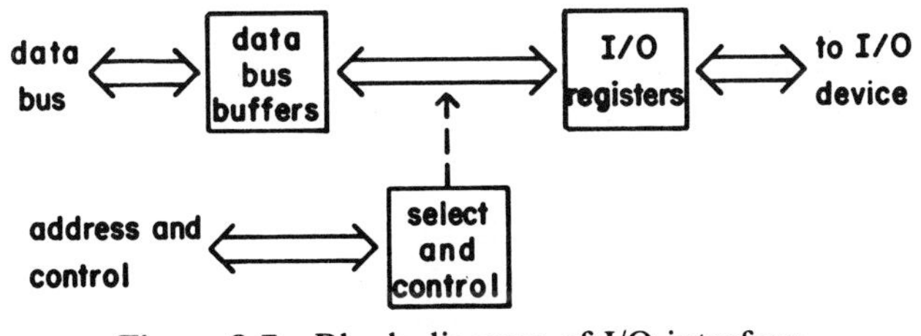

Figure 8.7 Block diagram of I/O interface

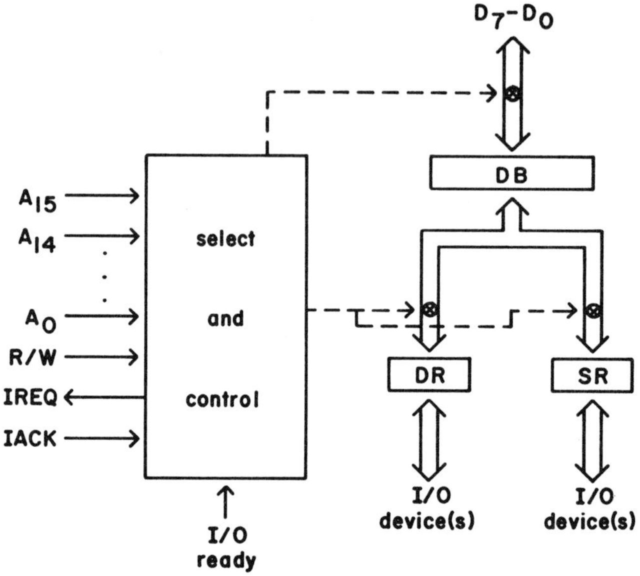

Figure 8.8 I/O interface—expanded diagram

With the above hardware available, the sequence of events occurring during an interrupt is then the following:

1. The I/O device sends an interrupt request through the I/O ready line which is transmitted on the IREQ line to the CPU.
2. The CPU, on the detection of an interrupt request, usually at the beginning of an instruction cycle, stores the status of the program presently in execution, sends back an acknowledge signal on the IACK line to start communication with the interface or I/O device.
3. Since there is usually more than one interface unit connected to the CPU, the CPU has to be able to know the source of the interrupt. This can be done in software by reading the status bits in SR in each of the interfaces in turn and testing for the setting of a particular bit. This method is known as *interrupt polling* and is adopted in a number of microprocessor systems.
4. The CPU makes use of the information read in from SR to branch to a particular location in memory to execute a specific interrupt service routine to achieve the desired I/O transfer operation.
5. Once the I/O transfer operation has been achieved, the stored status

of the program whose execution has been interrupted is restored to resume normal program execution.

The MC6820 peripheral interface adapter (PIA) given in Fig. 5.11 in Chapter 5 is designed in a similar fashion. The PIA, however, is actually two interfaces incorporated into one so that two sets of miltiplexed I/O devices can be connected to a CPU using one PIA. In addition, the number of interface address lines have been significantly reduced: there are three chip select lines CS0, CS1, and CS2, and the register select lines RS0 and RS1. This only means that an external circuit has to be added.

8.5 SERIAL I/O INTERFACES

Almost all data used in a computer system, both alphanumeric and numeric, are logically organized in parallel units. Consequently, the CPU, the main and secondary storage devices, and most of the commonly encountered I/O devices utilize parallel bits of information. On the other hand, certain devices like teletypewriters operate on serial data and the interfacing of remote I/O terminals to a computer also requires communication facilities which generally work on serial-bit data. Data formats for serial transmission are already very well standardized and include special bits for synchronization, control and hand-shaking (generation of status testing and acknowledge signals) in addition to the data bits. As an example, the format for the transmission of a character to and from a teletypewriter is shown below, where the sequence of 8 bits of information representing a 7-bit ASCII code and a parity bit is preceded by a START bit and followed by two STOP bits:

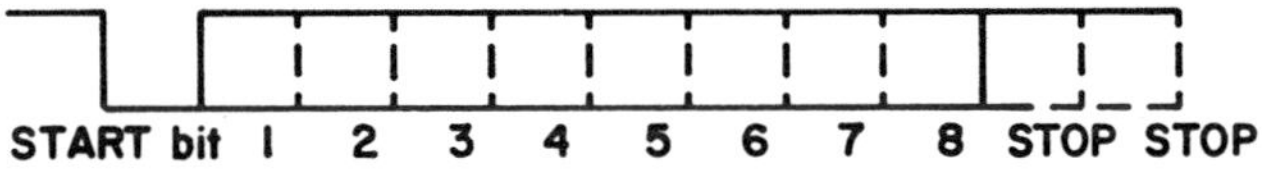

A listing of the set of ASCII codes can be found in the Appendix.

Interfacing the computer to parallel input devices is more or less straightforward. Each unit of information, generally a byte, is made available at the same time so that the interface is required to perform only a minimum amount of bookkeeping. If the input to or output from the computer is of a form different from that accepted by the I/O devices to which the interface is being connected, then transformation of incoming and outgoing signals must be done by the interface.

When the I/O device operates on serial data, the situation is much more complex and the interface involved must be able to perform signal trans-

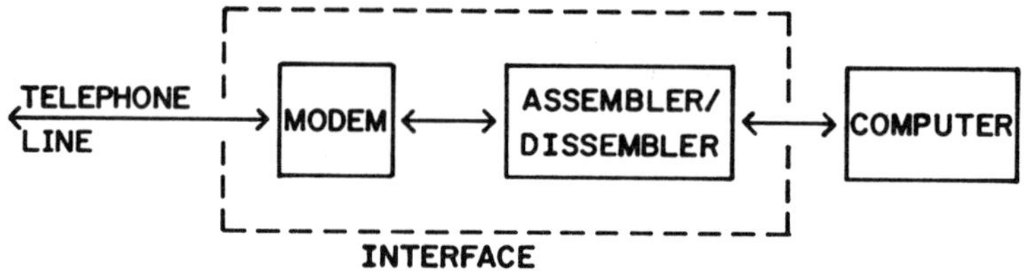

Figure 8.9 A serial I/O interface example

formation and also to assemble and dissemble data. An example of a serial I/O device is the telephone line commonly used in data communication among physically remote areas. The situation is shown in Fig. 8.9. Here, the modem (modulator-demodulator) is an acoustic coupler which is used to transform voltage signals into acoustic signals and vice versa. The assembler/dissembler assembles the serial input data into units, e.g., bytes, before they are transmitted to the computer and dissembles the parallel data from the computer into serial data for their output to the telephone line through the modem. This assembling and dissembling proc-

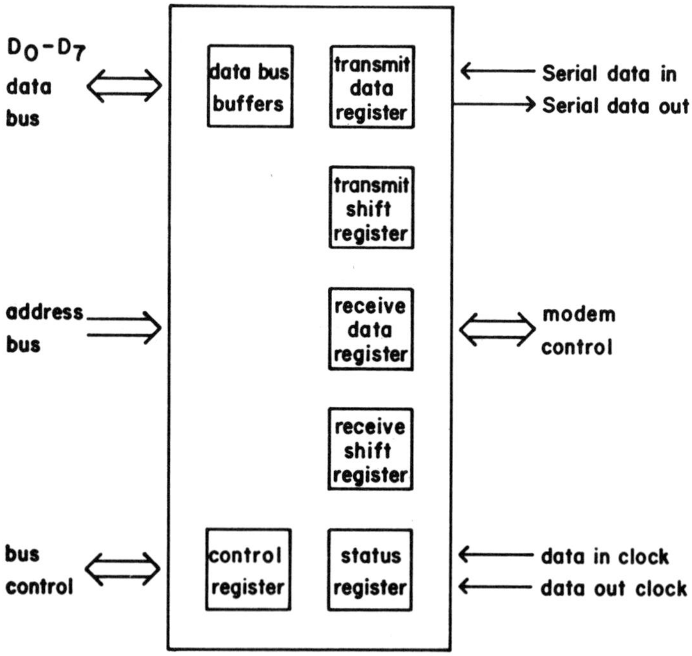

Figure 8.10 MC6850 ACIA block diagram

ess is typical of a serial I/O interface and usually involves the use of a counter as well as a number of registers which have to be present in the interfaces.

A block diagram of one such serial I/O interface unit, the Motorola MC6850 asynchronous communication interface adapter (ACIA), is given in Fig. 8.10. In addition to the various sets of bus interfacing and control lines, the figure also shows some of the more important registers residing in the interface unit: the data bus buffers, the data and shift registers for dissembling and transmission of data to the I/O device, the shift and data registers for the reception and assembling of data from the I/O device, and the control and status registers used for various control and bookkeeping operations. The two clock-input terminals are needed for the synchronization of the serial data input from and the output to the I/O device.

There is normally an additional operation that must be performed by a serial I/O interface. This is the generation and checking of parity bit information which is used to maintain a high degree of integrity in the data transmitted and received.

PROBLEMS

8.1 Describe the three different methods of I/O data transfer: programmed transfer, interrupt transfer, and DMA transfer. In each case base your description on the operations required to be performed by
(a) the CPU
(b) the interface unit.

8.2 Relate the components in the general I/O interface unit shown in Fig. 8.8 to those in the MC6820 PIA shown in Fig. 5.11.

8.3 Study the I/O instructions in the IBM 360 machine and their associated operations performed in the CPU and an I/O channel (interface).

8.4 Repeat Problem 8.3 for the Intel 8080 microcomputer.

8.5 Expand the instruction set in Table 7.1 to include an I/O instruction and to permit programmed transfer. How are the designs of the control unit discussed in Chapter 7 affected in each of the following cases:
(a) hardwired design,
(b) microprogrammed design?
Assume that each I/O interface is associated with two 1-byte buffer registers, one of them being the device status register and the other one the data register.

8.6 Repeat Problem 8.5 to permit single-level interrupt handling.

8.7 Repeat Problem 8.5 to permit DMA transfers.

8.8 Design an interface between a Boolean input of 0 or 1 and the CPU. Every time this input changes value in either direction, an interrupt request is to be made. The CPU input service routine will then read in the input value through the data line in the data bus corresponding to the most significant bit position.

8.9 If 8 Boolean inputs, each having its own interrupt line, are to be connected to the CPU, discuss how the use of 8 copies of the circuit designed in Problem 8.8 can be avoided.

8.10 Write a programmed data transfer loop in the assembly language of the DEC PDP-11 computer.

8.11 Assume that two I/O interfaces are to be daisy-chained together to a CPU. Each interface has the following facilities:

- an interrupt flag bit
- a ready flag bit
- an identifying bit (preset to 0 for one interface, and preset to 1 for the interface)
- an 8-bit data register

Complete the design of such an interface to permit daisy-chaining. You are only required to present a circuit diagram as your solution.

8.12 Assume that a M6820 PIA is interfaced to an 11-bit shift register to permit the assembling of serial input expressed as 11-bit ASCII codes.
The operation can be briefly described as follows: The START signal is used to indicate to the MPU that an 11-bit code is now coming in as serial input. The MPU now shifts the serial input into the shift register and then commences to assemble the character from the shift register. The occurrence of a parity error is indicated by the signal PARERR.

1. Draw a flowchart to describe in detail the operation of the program that will achieve the aforementioned functions (show all major program loops).
2. What timing consideration(s) has to be taken care of in this circuit?

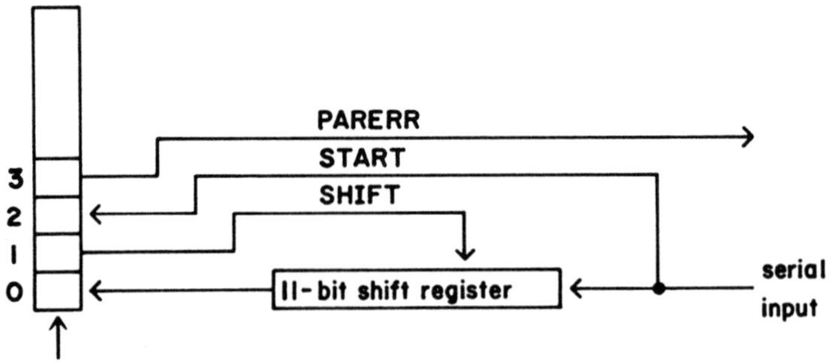

PIA data register A

REFERENCES

1. Flores, I. *Peripheral Devices*. Englewood Cliffs, N.J.: Prentice-Hall, 1973.
2. Hilburn, J. L., and Julich, P. M. *Microcomputers Micro Processors: Hardware, Software and Applications*. Englewood Cliffs, N.J.: Prentice-Hall, 1976.
3. *IBM System/360: Principles of Operation, GA22-6821*. IBM Corporation.
4. Kline, R. M. Digital Computer Design. Englewood Cliffs, N.J.: Prentice-Hall, 1977.
5. Lesea, A., and Zaks, R. *Microprocessor Interfacing Techniques*. Berkeley, CA, 1977.
6. Motorola, Inc. *M6800 Microcomputer System Design Data*. 1976.
7. Osborne, A. *An Introduction to Microcomputers, vol. I Basic Concepts*. Berkeley, CA: Adam Osborne and Associates, Inc., 1976.
8. Osborn, A. *6800 Programming for Logic Design*. Berkeley, CA: Adam Osborn and Associates, Inc., 1977.
9. Rony, P., Larsen, D., and Titus, J. *Bugbook III, Microcomputer Interfacing*. E & L Instruments, Inc., 1975.

Appendix A
7-bit ASCII Code Table

bit number 6 5 4 → 3 2 1 0		0 0 0	0 0 1	0 1 0	0 1 1	1 0 0	1 0 1	1 1 0	1 1 1
	binary (bit 6)	0	0	0	0	1	1	1	1
	binary (bit 5)	0	0	1	1	0	0	1	1
	binary (bit 4)	0	1	0	1	0	1	0	1
6 5 4 3 2 1 0	**hexadecimal**	0	1	2	3	4	5	6	7
0 0 0 0	0	NUL	DLE	SP	0	@	P	‘	p
0 0 0 1	1	SOH	DC1	!	1	A	Q	a	q
0 0 1 0	2	STX	DC2	″	2	B	R	b	r
0 0 1 1	3	ETX	DC3	#	3	C	S	c	s
0 1 0 0	4	EOT	DC4	$	4	D	T	d	t
0 1 0 1	5	ENQ	NAK	%	5	E	U	e	u
0 1 1 0	6	ACK	SYN	&	6	F	V	f	v
0 1 1 1	7	BEL	ETB	’	7	G	W	g	w
1 0 0 0	8	BS	CAN	(	8	H	X	h	x
1 0 0 1	9	HT	EM	)	9	I	Y	i	y
1 0 1 0	10	LF	SUB	*	:	J	Z	j	z
1 0 1 1	11	VT	ESC	+	;	K	[	k	{
1 1 0 0	12	FF	PS	,	<	L	\	l	|
1 1 0 1	13	CR	GS	−	=	M	]	m	}
1 1 1 0	14	SO	RS	.	>	N	Λ(↑)	n	~
1 1 1 1	15	SI	US	/	?	O	=(←)	o	DEL

INDEX